LIVING BY

Chemistry

First Edition

Unit 3: Weather Teacher Guide
Phase Changes and Behavior of Gases

Angelica M. Stacy
Professor of Chemistry
University of California at Berkeley

with

Janice A. Coonrod
Senior Writer and Developer

Jennifer Claesgens
Curriculum Developer

Key Curriculum Press

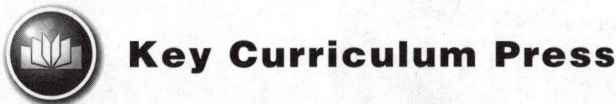

Editors	Ladie Malek, Jeffrey Dowling
Project Administrators	Elizabeth Ball, Rachel Merton, Janis Pope
Consulting Editors	Heather Dever, Joan Lewis, Andres Marti
Editorial Advisor	Casey FitzSimons
Production Editor	Andrew Jones
Editorial Production Supervisor	Kristin Ferraioli
Copyeditor	Mary Roybal
Senior Production Coordinator	Ann Rothenbuhler
Production Director	Christine Osborne
Text Designer	Roy Neuhaus Design
Compositor	Precision Graphics
Art Editor	Maya Melenchuk
Illustrators	Ken Cursoe, Greg Hargreaves, Tom Ward
Technical Artist	Precision Graphics
Photo Researcher	Laura Murray Productions
Cover Designer	Diana Ghermann
Prepress and Printer	Sheridan Books, Inc.
Textbook Product Manager	Tim Pope
Executive Editor	Josephine Noah
Publisher	Steven Rasmussen

This material is based upon work supported by the National Science Foundation under award number 9730634. Any opinions, findings, and conclusions or recommendations expressed in this publication are those of the author and do not necessarily reflect the views of the National Science Foundation.

© 2010 by Key Curriculum Press. All rights reserved.

No part of this publication may be reproduced, stored in a retrieval system, or transmitted, in any form or by any means, electronic, photocopying, recording, or otherwise, without the prior written permission of the publisher.

®Key Curriculum Press is a registered trademark of Key Curriculum Press.

The sciLINKS® service includes copyrighted materials and is owned and provided by the National Science Teachers Association. All rights reserved.

Photo Credits—ix: Ken Karp Photography; xii: Ken Karp Photography; xviii: Grafton Marshall Smith/Corbis; 166: NOAA/Handout/Reuters/Corbis

Key Curriculum Press
1150 65th Street
Emeryville, CA 94608
editorial@keypress.com
www.keypress.com

Printed in the United States of America
10 9 8 7 6 5 4 3 2 1   15 14 13 12 11 10 09
ISBN: 978-1-55953-990-6

Reviewers and Field Testers

Science Content Advisor

Dr. A. Truman Schwartz, *Macalester College (emeritus), St. Paul, MN*

Teaching and Content Reviewers

Scott Balicki
*Boston Latin School
Boston, MA*

Greg Banks
*Urban Science Academy
West Roxbury, MA*

Randy Cook
*Tri County High School
Howard City, MI*

Thomas Holme
*University of Wisconsin-Milwaukee
Milwaukee, WI*

Mark Klawiter
*Deerfield High School
Deerfield, WI*

Carri Polizzotti
*Marin Catholic High School
Larkspur, CA*

Matthew Vaughn
*Burlingame High School
Burlingame, CA*

Rebecca Williams
*Richland College
Dallas, TX*

Field Testers

Carol de Boer
*Amador Valley High School
Pleasanton, CA*

Wayne Brock
*Life Learning Academy
San Francisco, CA*

Janie Burkhalter,
*Coronado High School
Lubbock, TX*

Karen Chang
*The Calhoun School
New York, NY*

Elizabeth Christopher
*El Camino High School
Woodland, CA*

Mark Crown
*Gateway High School
San Francisco, CA*

Susan Edgar-Lee
*Hayward High School
Hayward, CA*
and
*Livermore High School
Livermore, CA*

Melissa Getz
*Tennyson High School
Hayward, CA*

Shannon J. Halkyard
*Stuart Hall High School
San Francisco, CA*

David Hodul
*De La Salle High School
Concord, CA*
and
*Bishop O'Dowd High School
Oakland, CA*

Field Testers (continued)

Kim D. Johnson
*Thurgood Marshall Academic
High School*
San Francisco, CA

Evy Kavaler
Berkeley High School
Berkeley, CA

Bruce Leach
Hill Country Christian School
Austin, TX

Tatiana Lim
Morse High School
San Diego, CA

Kathleen Markiewicz
Boston Latin School
Boston, MA

Steve Maskel
Hillsdale High School
San Mateo, CA

Mardi Mertens
Berkeley High School
Berkeley, CA

Nicole Nunes
*Thurgood Marshall Academic
High School*
San Francisco, CA
and
De La Salle High School
Concord, CA

Tracy A. Ostrom
Skyline High School
Oakland, CA

Pru Phillips
Crawfordsville High School
Crawfordsville, IN

Daniel Quach
Berkeley High School
Berkeley, CA

Carissa Romano
Hayward High School
Hayward, CA

Sally Rupert
Assets High School
Honolulu, HI

Geoff Ruth
Leadership High School
San Francisco, CA

Maureen Wiser
Emery Secondary School
Emeryville, CA

Audrey Yong
*Thurgood Marshall Academic
High School*
San Francisco, CA

Acknowledgments

A number of individuals joined the project as developers for various periods of time along the way to completing this work. Thanks go to these individuals for their contributions to the unit development: Karen Chang, David Hodul, Rebecca Krystyniak, Tatiana Lim, Jennifer Loeser, Evy Kavaler, Sari Paikoff, Sally Rupert, Geoff Ruth, Nicci Nunes, Gabriela Waschewski, and Daniel Quach.

David R. Dudley contributed original ideas and sketches for some of the wonderful cartoons interspersed throughout the book. His sketches provided a rich foundation for the art manuscript.

This work would not have been possible without the thoughtful feedback and great ideas from numerous teachers who field-tested early versions of the curriculum. Thanks go to these teachers and their students: Carol de Boer, Wayne Brock, Susan Edgar-Lee, Melissa Getz, David Hodul, Richard Kassissieh, Tatiana Lim, Evy Kavaler, Geoff Ruth, Nicci Nunes, Gabriela Waschewski, and Daniel Quach.

Dr. A. Truman Schwartz provided a thorough and detailed review of the manuscript. We appreciate his insights and chemistry expertise.

Ladie Malek and Jeffrey Dowling served as the developmental editors for the project, giving feedback and advice.

About the Author

Angelica Stacy is the author and lead developer of *Living By Chemistry*. In addition to her research and publications in materials, physical, and inorganic chemistry, she has distinguished herself as an outstanding educator, receiving numerous awards and honors in education and holding the President's Chair for Teaching at the University of California from 1993 to 1996. Dr. Stacy is a creative and enthusiastic educator whose interest in developing a high school chemistry curriculum arose out of her desire to help students attain a better understanding of the principles and concepts in chemistry. In 2005, the National Science Foundation named Dr. Stacy a Distinguished Teaching Scholar.

Contents
Unit 3 Weather: Phase Changes and Behavior of Gases

Introduction to the *Living By Chemistry* Program	viii
Features of the Teacher Guide	xiii
Content Coverage Chart	xvi
Introduction to Weather: Phase Changes and Behavior of Gases	xviii
Pacing Guides	xx

SECTION I Physically Changing Matter   1

Lesson 1 Weather or Not: *Weather Science*	2
Lesson 2 Raindrops Keep Falling: *Measuring Liquids*	11
Lesson 3 Having a Meltdown: *Density of Liquids and Solids*	20
Lesson 4 Hot Enough: *Thermometers*	29
Lesson 5 Absolute Zero: *Kelvin Scale*	38
Lesson 6 Sorry, Charlie: *Charles's Law*	47
Lesson 7 Front and Center: *Density, Temperature, and Fronts*	55

SECTION II Pressing Matter   64

Lesson 8 It's Sublime: *Gas Density*	65
Lesson 9 Air Force: *Air Pressure*	74
Lesson 10 Feeling Under Pressure: *Boyle's Law*	83
Lesson 11 Egg in a Bottle: *Gay-Lussac's Law*	92
Lesson 12 Be the Molecule: *Molecular View of Pressure*	104
Lesson 13 What Goes Up: *Combined Gas Law*	113
Lesson 14 Cloud in a Bottle: *High and Low Air Pressure*	120

SECTION III Concentrating Matter   127

Lesson 15 *n* Is for Number: *Pressure and Number Density*	128
Lesson 16 STP: *The Mole and Avogadro's Law*	138
Lesson 17 Take a Breath: *Ideal Gas Law*	146
Lesson 18 Feeling Humid: *Humidity, Condensation*	155
Lesson 19 Hurricane!: *Extreme Physical Change*	165
Lesson 20 Stormy Weather: *Unit Review*	173

Introduction to the *Living By Chemistry* Program

Living By Chemistry is a full-year high school chemistry curriculum that meets and exceeds state and national standards. It consists of six teacher guides, a student textbook, kits, and other print and online teaching resources. The teacher guides are central to the curriculum and provide detailed daily lesson plans. The textbook is accessible and highly visual, and follows the sequencing and flow of what happens in the classroom, directly supporting and reviewing the daily learning.

The curriculum consists of six units, each organized around a specific body of chemistry content and a theme that students can relate to. Most units consist of around 20 lessons of 45-minute duration, which can be combined for 90-minute block periods.

Unit 1	Alchemy	Matter, Atomic Structure, and Bonding	28 Lessons
Unit 2	Smells	Molecular Structure and Properties	22 Lessons
Unit 3	Weather	Phase Changes and Behavior of Gases	20 Lessons
Unit 4	Toxins	Stoichiometry, Solution Chemistry, and Acids and Bases	27 Lessons
Unit 5	Fire	Energy, Thermodynamics, and Oxidation-Reduction	20 Lessons
Unit 6	Showtime	Reversible Reactions and Chemical Equilibrium	8 Lessons

A Thematic Approach

A theme-based curriculum captures students' interest, helps them make connections, and improves retention of concepts. It also serves another purpose—it helps to ground the study of chemistry in the natural world and everyday life. Too often, students view chemistry as an inaccessible discipline centered around synthetic chemicals invented in a lab. In reality, chemical processes occur in our bodies and in the world around us all the time. Without most of these processes, life would not be possible. *Living By Chemistry* supports teachers in fostering students' wonder and curiosity about the world around them.

Science as Guided Inquiry

Living By Chemistry is the product of a decade of research and development in high school classrooms, focusing on optimizing student understanding of chemical principles. The curriculum was developed with the belief that science is best learned through first-hand experience and discussion with peers. Guided inquiry allows students to actively participate in, and become adept at, scientific processes and communication. These skills are vital to a student's further success in science as well as beneficial to other future pursuits.

The *Living By Chemistry* curriculum provides you with a student-centered lesson for each day. Students have opportunities to ask questions, make scientific observations, collect evidence, and formulate scientific hypotheses and explanations. In each lesson, students discover concepts and communicate ideas with peers and with the teacher. Formal definitions and formulas are frequently introduced *after* students have explored, scrutinized, and developed a concept, providing more effective instruction.

Thinking Like a Scientist

Using guided inquiry as a teaching tool promotes scientific reasoning, critical-thinking skills, and a greater understanding of the concepts. Students develop their own logical conclusions and discover chemistry concepts for themselves, rather than accepting and memorizing facts. The ultimate goal is to foster students who think like scientists and understand the nature of scientific practice. Students learn to study the natural world by asking questions, and proposing explanations based on evidence. They learn to reflect on their ideas and review their work and that of their peers, and to effectively communicate scientific concepts they have discovered.

Chemistry for All Students

Chemistry is at the core of many aspects of our daily lives. Now, more than ever, the world needs citizens who can make informed decisions about their health, the environment, energy use, nutrition, and safety. In addition, chemistry is a required course for a myriad of different career paths relating to science, engineering, health, and the environment. The *Living By Chemistry* curriculum helps you to promote scientific literacy and support all students in developing valuable skills that extend well beyond the classroom.

Sequenced for Understanding

The sequencing of topics in *Living By Chemistry* is purposeful and well-tested. The topics are ordered and presented in a way that optimizes understanding. In addition, the curriculum covers all necessary standards and concepts. (See the Content Coverage Chart on pages xvi–xvii.) You may be tempted to reorder topics or front-load detailed information when a topic is introduced—this is not necessary because *Living By Chemistry* is a spiraling curriculum in which topics are revisited in increasing depth throughout the course. Our experience confirms that by building a solid foundation and by scaffolding all the topics—including the mathematics—*Living By Chemistry* can help you prepare your students for even the most challenging topics.

A Typical Day

Start With Student Understanding

Students come into any class with prior knowledge, assumptions, and misconceptions. When students make sense of new evidence and revise their thinking to accommodate it, they build true understanding. *Living By Chemistry* supports this process by allowing students to build their understanding based on experiences and including discussion questions specifically designed to challenge them to share their reasoning. Lessons are constructed to lead students to a more complete understanding.

The first step in this learning process is to discover students' current understanding of a concept or subject, through discussion or written questioning. Students' understanding is likely to be flawed or incomplete, so it is vital to create a safe atmosphere in the classroom for the sharing of these ideas. To do this, the *Living By Chemistry* curriculum uses a modified version of the 5Es model of teaching and learning. Originally created by a team led by Principal Investigator Rodger Bybee at the Biological Sciences Curriculum Study (BSCS), the 5Es describe different stages of a learning sequence: Engage, Explore, Explain, Elaborate, and Evaluate.

Lessons That Promote Understanding

Engage: At the start of each class, students are immediately engaged in a brief warm-up exercise, called a ChemCatalyst, that focuses on the main goal of the lesson. The purpose of the ChemCatalyst is to determine students' prior knowledge on the subject and encourage participation. In most cases, you can listen to student ideas and ask for explanations without judgment or correction.

ChemCatalyst

1. If you were to drop a spoonful of salt, NaCl, into a glass of water, what would happen?
2. If you were to drop a gold ring into a glass of water, what would happen?
3. What do you think is different about the atoms of these two substances? Why do you suppose the gold atoms don't break apart?

Explore: For the next 15 to 20 minutes of class, students explore the key chemistry topics covered that day. Depending on the lesson type, they might solve problems, analyze data, perform a laboratory experiment, build models, or complete a card sort activity. They might also watch a brief demonstration or a computer simulation and try to provide explanations for their observations. This is a chance for students to think and build their own understanding, and, most importantly, to support their ideas with evidence. Generally, students work collaboratively in small groups or pairs. During this portion of the lesson, you can circulate from group to group, offering guidance, asking questions, and helping students to refine their ideas.

Explain and Elaborate: A teacher-led discussion follows the Explore portion of the class. It allows students to connect their conceptual understanding from the lesson with new chemistry concepts, ideas, tools, or definitions that make up the learning objectives. Sample discussion questions are provided for you, along with summaries of the key points to be covered.

Evaluate: At the end of class, a final Check-in question provides both you and students with a quick assessment of their grasp of the day's main concepts.

> **Check-in**
>
> Predict whether $MgSO_4(aq)$, commonly known as Epsom salts, will conduct electricity. State your reasoning.

Homework: Each lesson is accompanied by a reading assignment in the student textbook. Usually this amounts to about two or three pages of reading that reviews and reinforces what was learned in class, followed by exercises. The reading includes diagrams, photos, worked examples, and real-world connections. *Living By Chemistry* is designed so that each reading *follows* its corresponding lesson. This reinforces the concepts developed in class, and therefore optimizes the effectiveness of both the classroom experience and the reading component.

Other Resources

Kits

Specialty items specifically designed for *Living By Chemistry,* such as card decks and molecular modeling sets, are provided in kits. Kits contain enough materials for classes of up to 32 students. Replacements are available for purchase. Additional materials needed for teaching each unit, including lab equipment and chemical supplies, are listed online at www.keypress.com/keyonline.

Lesson Presentations

Lesson presentations are also available in Microsoft® PowerPoint® format. For these and other resources, please visit www.keypress.com/keyonline.

Topic: Isotopes
Visit: www.SciLinks.org
Web code: KEY-114

SciLinks

Our partnership with SciLinks allows you to access the best resources available on the Internet for specific topics. Links take you to websites selected by the National Science Teachers Association.

Tips and Classroom Management Strategies

Share the Approach

For some students, the *Living By Chemistry* approach will be new and different from their prior science classroom experiences. Let your students know that they will be taking an active role in their own learning, and that seeking the "right" answer will not always be the best strategy. Let them know that you will often be more interested in *why* they think something rather than whether or not what they think is scientifically correct. This will help to establish an open and safe atmosphere where students can share ideas and build their knowledge of chemistry.

Begin With the Basics

Living By Chemistry begins with the basics. We don't assume that students come into your classroom with any exposure to chemistry or its concepts. This scaffolded approach means that students new to chemistry are not at an instant disadvantage.

At the same time, students who are acquainted with chemistry vocabulary and principles have an opportunity to clarify and expand their understanding. You'll find that while lessons may start simply, they build to embrace sophisticated concepts and principles. Although *Living By Chemistry* might appear to start more slowly than other courses you've seen, all students benefit in the long run by developing a deeper and more comprehensive understanding.

Adjust Pacing to Match the Class

The flexibility of the *Living By Chemistry* curriculum allows you to adjust pacing to meet the needs of each class. For example, for an honors class the pacing and expectations can be increased. Some lessons can be completed as homework, more projects and Internet research can be assigned, and concepts can be pursued to more sophisticated levels. Other classes may work at a slower pace, with more time spent discussing, practicing, and clarifying concepts. For these classes, you may wish to occasionally spend an extra day on a lesson. For double-periods, two lessons can often be easily merged into one. (See Pacing Guides on page xx.)

Cooperative Learning

In the *Living By Chemistry* curriculum, students work cooperatively in groups, gaining expertise by articulating ideas and communicating concepts. This is one of the best ways to become proficient in a discipline. Students benefit from seeing a variety of approaches to a single problem, and they use their group as a sounding board for their ideas, rather than struggling alone or having to ask you.

You might assign new groups at the start of each unit. Some teachers select groups based on students' strengths or other criteria, while some teachers prefer to assign groups randomly.

One strategy that lends to effective group work is a fair division of tasks among the group members. Each member could be assigned a role, such as spokesperson, recorder, equipment and lab safety person, or facilitator. In this way, the work is shared, everyone is involved, and everyone has an opportunity to practice communication, leadership, and responsibility.

Success for More Students

Extensive field-testing and research over the past decade have shown that students of all ability levels perform better with *Living By Chemistry*. One of the gratifying comments that we hear frequently from teachers is that they find many more students participating and engaged, including those who were not participating previously. We hope your experience with this curriculum meets and exceeds your expectations, and we welcome your feedback. Contact us at editorial@keypress.com.

Features of the Teacher Guide

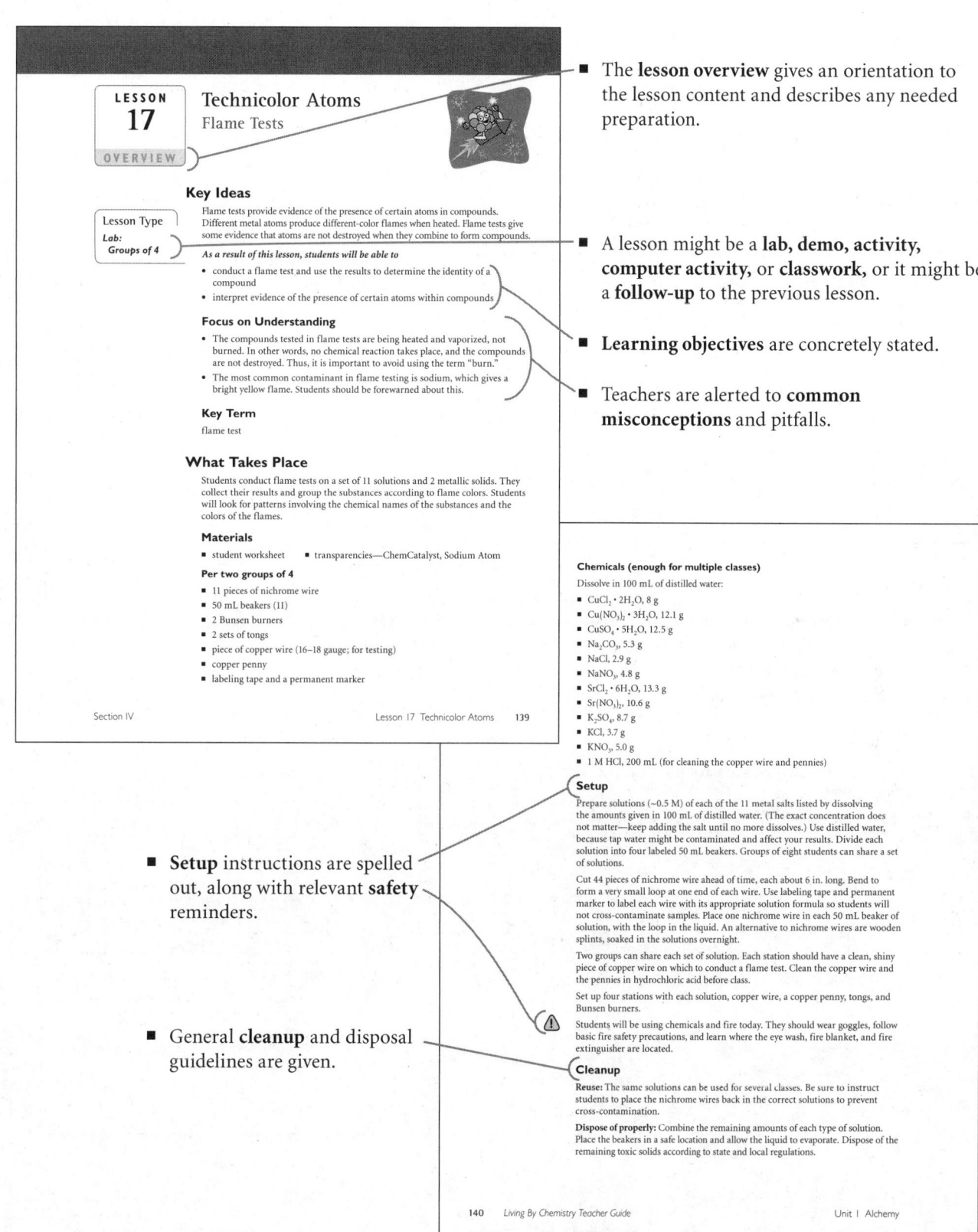

- The **lesson overview** gives an orientation to the lesson content and describes any needed preparation.

- A lesson might be a **lab, demo, activity, computer activity,** or **classwork,** or it might be a **follow-up** to the previous lesson.

- **Learning objectives** are concretely stated.

- Teachers are alerted to **common misconceptions** and pitfalls.

- **Setup** instructions are spelled out, along with relevant **safety** reminders.

- General **cleanup** and disposal guidelines are given.

xiii

LESSON 25 LESSON GUIDE

You Light Up My Life
Classifying Substances

Engage (5 minutes) ——— ■ Every lesson begins with an **Engage** section.

Key Question: How can substances be sorted into general categories?

ChemCatalyst ——— ■ Students warm up with the **ChemCatalyst**, designed to engage them as well as help the teacher determine prior knowledge and misconceptions.

1. If you were to drop a spoonful of salt, NaCl, into a glass of water, what would happen?
2. If you were to drop a gold ring into a glass of water, what would happen?
3. What do you think is different about the atoms of these two substances? Why do you suppose the gold atoms don't break apart?

Sample Answers: 1. The salt would sink to the bottom of the glass and would begin to dissolve. Eventually it would all mix into the water. Some students might say the salt will "disappear."
2. The gold ring also will sink to the bottom of the glass, but then nothing will happen to it.
3. Students might hypothesize that the salt can break apart because it is made from two different elements, or that gold atoms have a stronger connection with one another.

Discuss the ChemCatalyst
→ Assist students in verbalizing their thoughts about the properties of gold and about dissolving in particular.

Sample Questions ——— ■ **Sample questions** help you guide the discussion.
- What do you think is happening to the atoms when salt dissolves?
- Why do you think gold doesn't dissolve?
- How are the atoms of these two substances different?
- Why do you think some solids dissolve while others don't?

Explore (20 minutes) ——— ■ As part of the **Explore** section, students work together to discover and make sense of new chemistry content.

Introduce the Lab
→ Briefly define the terms *dissolving* and *conductivity*.

> **Dissolve:** To disperse evenly into another substance. For example, a solid can dissolve in a liquid.
> **Conductivity:** A property that describes how well a substance transmits electricity.

Section V — Lesson 25 You Light Up My Life — 211

■ The **reduced student worksheet** is shown for reference. The blackline master is in the *Teaching and Classroom Masters* book.

■ Sample **answers** are provided on the reduced pages.

LESSON 18 CLASSWORK

Life on the Edge
Valence and Core Electrons

Name _____ Date _____ Period _____

Purpose
To discover the arrangements of electrons within atoms.

Instructions
Complete the table on the handout, filling in the missing atoms. Then answer the questions.

1. How does the number of electrons change as you move from left to right across a period?
   The number of electrons increases by one as you move from left to right, from one element to the next.
2. What do all the atoms of Group 1A elements have in common?
   They all have one electron in the outermost shell.
3. List three things that all the atoms of the elements in period 3 have in common.
   1. They all have three shells. 2. There are two electrons in the first shell. 3. There are eight electrons in the second shell.
4. Which atoms have two electrons in the first shell and eight electrons in the second shell?
   neon, Ne, and all the elements in Periods 3 and 4
5. What happens to the electron count and the number of shells when you move from neon, Ne, to sodium, Na?
   A new shell is added. Also, one electron is added.
6. How many shells of electrons does rubidium, Rb, have? How many electrons are in the outermost shell? Draw a shell model of a rubidium atom.
   Rubidium, Rb, is in Group 1A at the beginning of period 5. It has five shells, with one electron in the outermost shell. Rubidium is the element with the next higher atomic number after krypton, Kr, so it looks just like krypton with a fifth shell added and one electron in the fifth shell.
7. Draw a shell model of an atom with two shells and six electrons. What element is this? How many electrons are in the outermost shell?
   The element is carbon. It has four electrons in the outermost shell.
8. Draw a shell model of an atom with three shells and two electrons in the outermost shell. How many total electrons does this atom have? What element is this?
   There are a total of 12 electrons: The first shell has 2, the second shell has 8, and the third shell has 2. The element is magnesium.

Living By Chemistry Teaching and Classroom Masters: Units 1–3
© 2010 Key Curriculum Press
Unit 1 Alchemy 67
Lesson 18 • Worksheet

Section IV — Lesson 18 Life on the Edge — 151

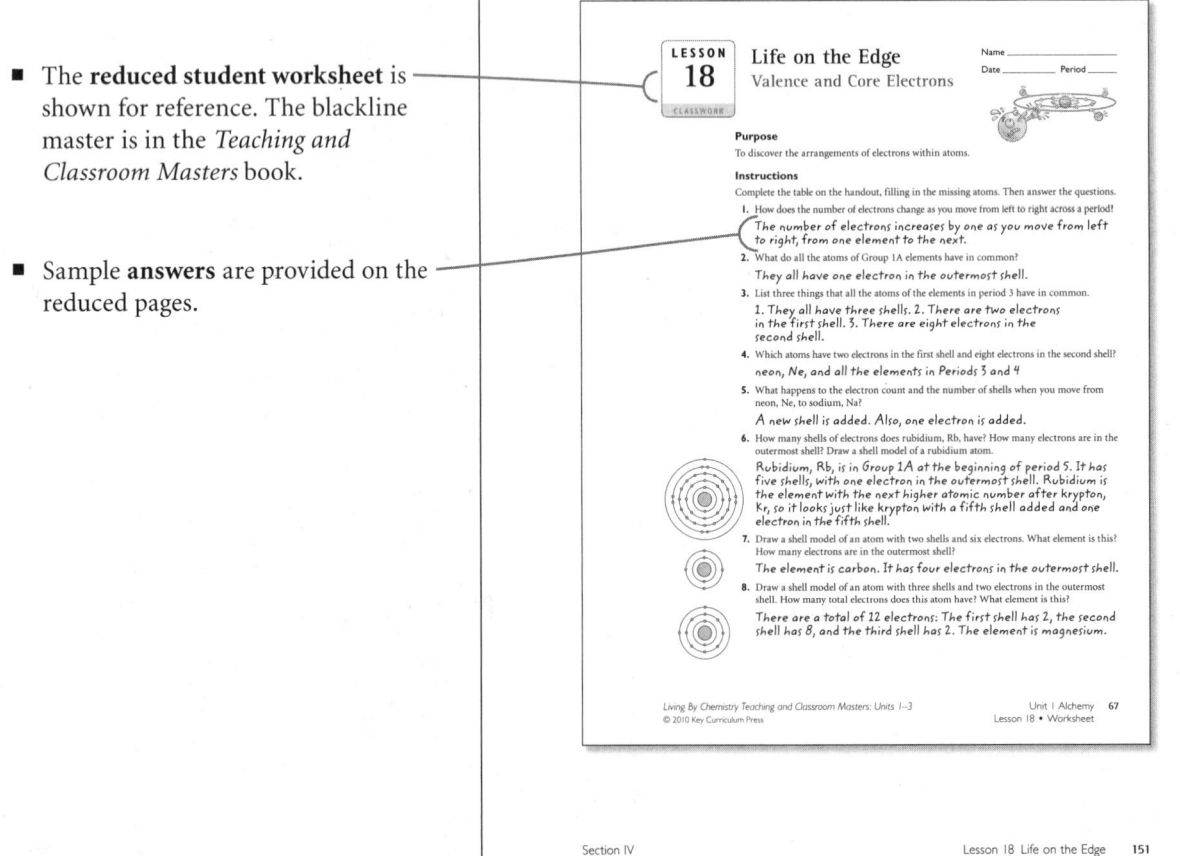

xiv

Explain and Elaborate (15 minutes)

Share Students' Generalizations about Conductivity

➡ Summarize students' generalizations and write them on the board as students share them. Ask for consensus on the statements as you accept them.

Sample Questions

- What generalizations can you make about the substances that did not light up the bulb?
- What generalizations can you make about the substances that did light up the bulb?
- Based on your data, why do you think the sports drink lit up the bulb when dissolved?

Key Points

Generalizations about substances that *do not* light up the bulb:
- Compounds made up of C, H, and O atoms do not conduct electricity.
- Compounds made up entirely of nonmetals do not light up the bulb.
- Compounds made up of a combination of metals and nonmetals do not light up the bulb when they are in their solid form.

Generalizations about substances that *do* light up the bulb:
- Everything that lights up the bulb has a metal atom in it.
- Compounds made of metal and nonmetal atoms, such as salts, light up the bulb when they are dissolved in water. (The sports drink is a solution of water, various salts, sugar, and a dye.)
- Metal solids light up the bulb.

Analyze the Results

(T) ➡ Use the transparency Solubility and Conductivity to summarize the results. Ask students to assist you with filling in the substances at the bottom of the chart.
➡ After sorting the substances, introduce the terms *soluble* and *insoluble*. Label the appropriate boxes with these terms.

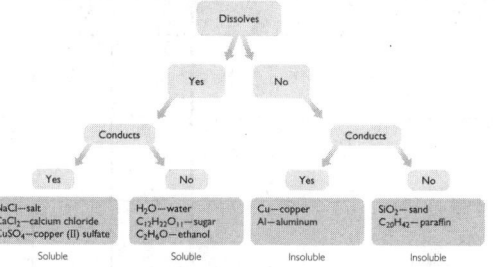

Sample Questions

- What statement can you make about ionic compounds and dissolving?
- What generalizations can you make about substances made up entirely of metal atoms?
- What generalizations can you make about substances made up entirely of nonmetal atoms?

Key Point

We can place all the substances tested into one of the four categories. A substance that dissolves in another substance is said to be "soluble" in that substance. A substance that does not dissolve in another substance is said to be "insoluble" in that substance. Once we find out whether a substance dissolves in water, we can further sort according to whether the dissolved substance conducts electricity. In the next lesson we will take a closer look at these categories in order to figure out what else the substances have in common, or what might account for their common properties.

> **Soluble:** Describes a substance that is capable of being dissolved in another substance.
> **Insoluble:** Describes a substance that is incapable of being dissolved in another substance.

Notice that substances that conduct electricity are either solid metals or ionic compounds dissolved in water. Substances made entirely of nonmetal atoms, such as sugar, do not conduct electricity.

Note: Some substances are considered slightly soluble. More discussion of solubility will take place in Unit 4: Toxins.

Wrap-up

Key Question: How can substances be sorted into general categories?

- Not all substances dissolve in water.
- Not all substances conduct electricity.
- Solid metals and metal–nonmetal compounds dissolved in water conduct electricity.

Evaluate (5 minutes)

Check-in

Predict whether $MgSO_4(aq)$, commonly known as Epsom salts, will conduct electricity. State your reasoning.

Answer: $MgSO_4$ dissolves in water and it contains both metal and nonmetal atoms so it will conduct electricity.

Homework

Assign the reading and exercises for Alchemy Lesson 25 in the student text.

Content Coverage Chart

Concepts are introduced and then reinforced, often across different units. Coverage usually consists of an introduction (I), then practice (P), and finally teaching to mastery (M).

Content	Alchemy I	Alchemy II	Alchemy III	Alchemy IV	Alchemy V	Smells I	Smells II	Smells III	Smells IV
Atomic and Molecular Structure			I	P	P	P	P	P	M
Chemical Bonds					I	P	P	P	P
Conservation of Mass and Stoichiometry	I	P	P						
Gases and Their Properties									
Acids and Bases		I				P			
Solutions		I							
Energy/Chemical Thermodynamics									
Reaction Rates						I			
Chemical Equilibrium									
Organic Chemistry and Biochemistry						I	P	P	P
Nuclear Processes			I/P						
Investigation and Experimentation	I	P	P	P	P	P	P	P	P

Unit and Section														
Weather			Toxins					Fire				Showtime	
I	II	III	I	II	III	IV	V	I	II	III	IV	I	II
											M		
			P	P	P	P	P	P	P	P	M		
I	P	M											
						P						P	M
					P	P	P				P	P	M
	I							P	P	P	P	M	
											P	P	P
												I	P
P	P	P	P	P	P	P	P	P	P	P	P	P	M

Introduction to Weather:
Phase Changes and Behavior of Gases

Weather as Context

Most of the weather phenomena we experience are a result of physical changes in matter. Differences in density cause the movement of air we experience as wind. Heat from the sun causes water to evaporate and rise into the atmosphere. Water vapor is moved about by temperature and pressure differentials in the air, only to change phase again and become some form of precipitation. To understand weather, we need to understand phase changes and the interplays of variables such as pressure, temperature, relative humidity, and density.

Content Driven by Context

There are notable advantages to using weather as a context. Students have probably constructed personal explanations for the causes of various weather phenomena they experience, from fog to hurricanes. Building on these first-hand experiences is a motivating way to study a topic as invisible and abstract as gas behavior.

Throughout the unit, students search for patterns among the weather variables. Understanding and predicting the weather is an intricate science, but students relish this opportunity to become amateur meteorologists and quickly gain some expertise.

Chemists study gas behavior by examining samples in closed containers. In contrast, our atmosphere is a constantly changing open system. To address this problem, in this unit we focus on air masses, which have uniform conditions of temperature and pressure. We also relate the study of specific volumes of gas or amounts of gas under controlled conditions to understanding what is happening in less controlled conditions. Ultimately, understanding the gas laws helps students to explain weather phenomena.

Mathematical Relationships

The gas laws involve mathematics that can be challenging for some students. In addition to algebra involving multiple variables, students face equations that may seem unfamiliar if they are used to solving for only x and y. Luckily, most of the mathematical relationships in chemistry are proportional ones, and they become more accessible to students when this feature is pointed out. Lessons in Unit 1: Alchemy have laid the groundwork by introducing proportional relationships and proportionality constants such as density. As they work with these proportional relationships, students begin to truly grasp the big picture and become more confident with algebraic relationships.

In this unit, we have taken special care with how the mole is introduced. Students regularly have difficulty with this concept, usually with grasping its magnitude and what it concretely represents. For this reason, we wait to use scientific notation until Unit 4: Toxins. The lessons in this unit make the mole as concrete as possible for students so that they are fully prepared for the stoichiometry in the next three units.

Math Spotlights on solving equations, dimensional analysis, signification digits, and other math topics are referenced where appropriate in the readings and examples in the student textbook. Found in the back of the student textbook, Math Spotlights provide additional review and practice on common math topics for your students.

The Use of Models to Explain Molecular Behavior

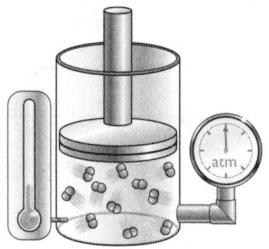

Understanding pressure, temperature, and volume changes of a gas hinges on understanding kinetic molecular theory. In order to succeed with the concepts in this unit, students need to be able to visualize gas molecules. To that end, the activities and the student textbook regularly help and encourage students to visualize gas behavior in this way. In addition, students explore molecular motion with computer simulations that help them visualize pressure, volume, and temperature changes.

Building Understanding

So far, students have constructed a solid foundation in chemistry by studying atomic and molecular structure and properties. In this unit, they build upon this foundation by studying gas laws, phase changes, temperature scales, and the concept of the mole. Topics in this unit are introduced in a deliberate sequence with the content and math carefully scaffolded.

Section I focuses on physical changes to matter, more specifically changes in phase, volume, and density of matter in response to changes in temperature. The proportional relationship between volume and temperature of a gas is introduced along with the Kelvin scale.

In **Section II,** students explore gas pressure qualitatively, quantitatively, and from a particulate viewpoint. By focusing on high- and low-pressure systems in the atmosphere, students learn to explain macroscopic observations of matter in terms of the behavior of particles. Lessons explore the relationships among volume, temperature, and pressure of gases.

Section III explores the relationship between the number of particles, n, and all the previously mentioned variables. This section introduces the mole, setting up a firm foundation for stoichiometry in the next unit. Exploring the number of particles allows us to investigate humidity, providing explanations for weather phenomena such as fog, dew, rain, and snow. Finally, the many variables are brought together in a lesson on hurricanes.

Pacing Guides

Standard Schedule

Day	Suggested Plan
1	Lesson 1
2	Lesson 2
3	Lesson 3
4	Lesson 4
5	Lesson 5
6	Lesson 6
7	Lesson 7, Section I Review, Project (optional)
8	Section I Quiz, Lesson 8
9	Lesson 9
10	Lesson 10
11	Lesson 11
12	Lesson 11 (cont.)

Day	Suggested Plan
13	Lesson 12
14	Lesson 13
15	Lesson 14, Section II Review, Project (optional)
16	Section II Quiz, Lesson 15
17	Lesson 16
18	Lesson 17
19	Lesson 18
20	Lesson 19
21	Section III Review, Project (optional)
22	Section III Quiz, Lesson 20
23	Unit Review, Lab Exam (optional)
24	Unit Exam

Block Schedule

Day	Suggested Plan
1	Lessons 1 and 2
2	Lessons 3 and 4
3	Lessons 5 and 6
4	Lesson 7, Section I Review, Project (optional)
5	Section I Quiz, Lesson 8 and 9
6	Lessons 9 (cont.) and 10
7	Lesson 11

Day	Suggested Plan
8	Lesson 12 and 13
9	Lesson 14, Section II Review, Project (optional)
10	Section II Quiz, Lessons 15 and 16
11	Lessons 17 and 18
12	Lesson 19, Section III Review, Project (optional)
13	Section III Quiz, Lesson 20, Unit Review
14	Unit Exam, Lab Exam (optional)

I Physically Changing Matter

Section I of the Weather unit introduces students to the physical changes of matter and explores how those changes are monitored, measured, and tracked. In the first lesson, students explore the information typically found on weather maps. They look for correlations among the location of the jet stream, clouds, precipitation, temperature, high and low pressure, and warm and cold fronts. In Lesson 2, students are introduced to the first of many proportional relationships when they investigate why rainfall is measured in units of height rather than volume. Students explore phase change behavior in Lesson 3, one of several lessons in the unit that highlight the interactions among phase, density, temperature, and volume. In Lesson 4, students construct simple liquid and gas thermometers. Lesson 5 introduces students to the Kelvin scale and the kinetic theory of gases. Lesson 6 explores the effect of temperature changes on gases and quantifies the relationship between gas volume and temperature. In Lesson 7, students apply what they have learned about gas behavior to the topics of weather fronts, cold and warm air masses, and the formation of storms and precipitation.

In this section, students will learn

- how to interpret basic weather maps
- about basic proportional relationships
- about thermometers and temperature scales
- the kinetic theory of gases
- phase change behavior
- Charles's law relating the volume and temperature of a gas

LESSON 1 OVERVIEW

Weather or Not
Weather Science

Lesson Type
Activity:
Whole-class or Groups of 4

Key Ideas

Weather on our planet is a result of the dynamic interplay among Earth, the atmosphere, water, and the Sun. Meteorologists monitor moisture, air pressure, temperature, storm fronts, and the jet stream in order to predict storms and precipitation. Weather science is intimately related to the chemistry of phase changes and gas laws.

As a result of this lesson, students will be able to

- explain the phenomenon of weather in general terms
- list the variables meteorologists study or measure in order to predict the weather
- describe the basic components of a weather map

Key Terms

weather   physical changes   phase change

What Takes Place

This lesson introduces the Weather unit. Students examine data on weather maps to begin to understand weather forecasting. Data on the jet stream, temperature, cloud cover, weather fronts, precipitation, and air pressure are provided on individual maps, each on a separate color transparency. These transparencies can be overlaid during the class discussion so that students can look for relationships among the weather variables. Students also are introduced to some basic weather vocabulary that they will be using throughout this unit. (*Note:* We recommend that you obtain enough color transparencies so that students can work in groups on this activity, each with their own set; or photocopy the handout onto a transparency for each group and cut it apart so they can have their own black and white set. Alternatively, the lesson can be run using one set of color transparencies and an overhead projector. To do this, be sure to give students the handout—Weather Variables, so they can hypothesize about the relationships between variables.)

SciLinks NSTA
Topic: Weather
Visit: www.SciLinks.org
Web code: KEY-301

Materials

- student worksheet (optional)
- transparency— ChemCatalyst
- handout—Weather Variables

Per group of 4 (or one class set)

- 7 weather map color transparencies (Jet Stream, Temperature Highs, Cloud Cover, Fronts, Precipitation, Air Pressure, and blank base map) and a sheet of white paper (recommended; or one class set and an overhead projector)

LESSON 1 GUIDE

Weather or Not
Weather Science

Engage (5 minutes)

Key Question: What causes the weather?

ChemCatalyst

The table gives the current weather conditions in Miami, Florida (shown on the map with a star). Predict the weather for later today. Indicate whether you think the current conditions will increase, decrease, or stay the same. Explain your reasoning. Weather map for October 27, 2008:

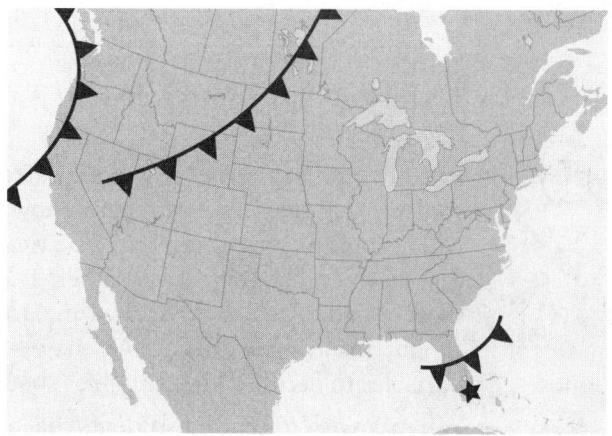

Current Conditions at 1:30 P.M. in Miami, Florida

Temperature	82 °F/27.8 °C
Pressure	29.95 in.Hg, falling
Fronts	Cold front to pass through today
Conditions	Mostly cloudy, wind gusts up to 30 mph NW
Humidity	71%

Sample Answer: Mainly cloudy and humid. High around 86 °F with showers and a thunderstorm today. Showers, thunderstorms, and wind tonight with a low of 72 °F. Clearing and sunny on Wednesday after morning showers. Reasons that students offer will vary.

Discuss the ChemCatalyst

Sample Questions

- What do you think is the most useful information in the weather data? Explain your reasoning.
- What was your forecast? What did you base it on?
- What things do you need to know about in order to forecast the weather?
- What do you think causes weather?

Explore (20 minutes)

Introduce the Activity

- Define weather. Point out that it is different from climate, which describes the average weather for a region over long periods of time.

> **Weather:** The state of the atmosphere in a region over a short period of time. Weather is the result of the interaction among Earth, the atmosphere, water, and the Sun. It refers to clouds, winds, temperature, and rainfall or snowfall.

- Tell students that a weather forecast is a prediction made by a meteorologist about the weather.
- Pass out a set of weather map transparencies to each group of four students. Tell students that these transparencies show weather data for the same day. They will be looking for patterns in these weather maps in order to better understand the variables that affect the weather. They can overlay the maps on a white sheet of paper, two or three at a time, to compare different variables. If you are conducting a whole class activity have students refer to the handout—Weather Variables to decide which variables should be compared.
- You may wish to remind students that a variable is a quantity or characteristic that can change or vary.

Guide the Activity

- If you are conducting a whole-class activity with one set of transparencies, start by displaying each transparency over the base map so students can answer Question 1. For Question 2, overlay Precipitation, Cloud Cover, and Temperature highs. Next, ask students to hypothesize about which weather variables may be related. Overlay the appropriate transparencies so students can check their hypotheses. You might want to do part 2 of the worksheet as a class discussion as well.

LESSON 1 ACTIVITY

Weather or Not
Weather Science

Name _____
Date _____ Period _____

Purpose

To explore weather maps in order to understand what causes the weather.

Instructions

These weather maps are from a day in early March. Use them to look for relationships between and among the factors that affect the weather.

Part 1: Weather Variables

1. Describe the type of information is given on each of the six weather maps?
    Temperature Highs: *Highs to the nearest 10 °F.*
    Cloud Cover: *Location of clouds*
    Precipitation: *Rain and snowfall*
    Jet Stream: *not sure—some kind of trail*
    Fronts: *not sure; warm and cold fronts*
    Air Pressure: *not sure; areas of high and low pressure*

2. Describe the weather according to your maps in the locations listed here. Include the temperature, cloud cover, and precipitation.

    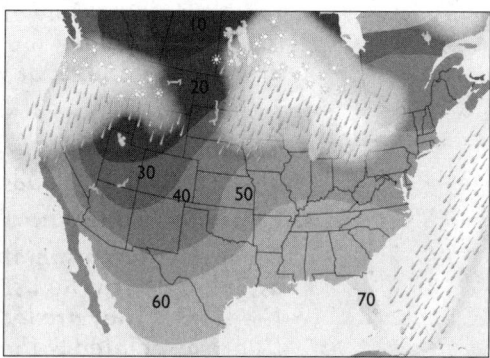

    a. Pacific Northwest

    some cloudiness leaving the area, mild temperatures ranging widely from 20 °F to 60 °F; no precipitation except for the farthest northeast corner of California

    b. Southeastern United States

    clear skies, no precipitation, warm, in the 60s and 70s

    c. Great Lakes region

    cloudy, rainy, and slightly chilly

3. Place the weather map transparencies on top of each other to compare more than one variable. List at least six patterns you notice.

200 Unit 3 Weather
Lesson 1 • Worksheet

Living By Chemistry Teaching and Classroom Masters: Units 1–3
© 2010 Key Curriculum Press

Section 1   Lesson 1 Weather or Not   5

Example: Where there are H's there are no clouds.

Sample answers:

- *Where there is precipitation there are also fronts.*
- *The temperature pattern follows the jet stream.*
- *The L's are located near where the blue and red fronts touch each other.*
- *The snow is located in the north, where it is much colder.*
- *There is precipitation where the L's are located.*

Part 2: Analysis

1. In the jet stream, air moves swiftly generally from west to east at high altitudes across the United States. How is the weather north of the jet stream different from the weather south of the jet stream?

   The weather north of the jet stream is quite a bit colder than the weather south of the jet stream.

2. Name a state where it may rain tomorrow. Provide your reasoning.

   Sample answer: Anywhere there are clouds or rain there may be rain the next day as well. The fronts seem to be associated with rain. Weather moves generally from west to east (the direction of the jet stream), so any state that is directly east of the rain on the map will get precipitation. Parts of Wisconsin, Minnesota, perhaps southern Nevada or western Utah, parts of Montana, Idaho, or Wyoming may experience rain.

3. Describe in your own words what you think a front is.

   Sample answer: Maybe it is where a storm is, or a mass of moist air, or an area of clouds.

4. What specific weather patterns are associated with the H's and L's on the map?

   Clear skies are associated with the H's. Stormy weather is associated with the L's. The L's are also closely associated with the fronts. They are located where the blue and red lines meet. L's are associated with clouds and with precipitation. Neither appears to be associated with any particular temperature.

5. **Making Sense** Use the weather maps of the United States to forecast the weather for your hometown or a location specified by your teacher. List the current conditions along with your forecast for the next day.

   Answers will vary from location to location.

6. **If You Finish Early** Which of the maps can you use to decide whether it is a sunny day in a certain place? Explain your thinking.

   The cloud cover maps help you decide whether it is a sunny day somewhere. The air pressure maps also help you to decide if it is sunny. Wherever there are H's there are clear skies. Precipitation maps can tell you where it is certainly not sunny.

WEATHER VARIABLES

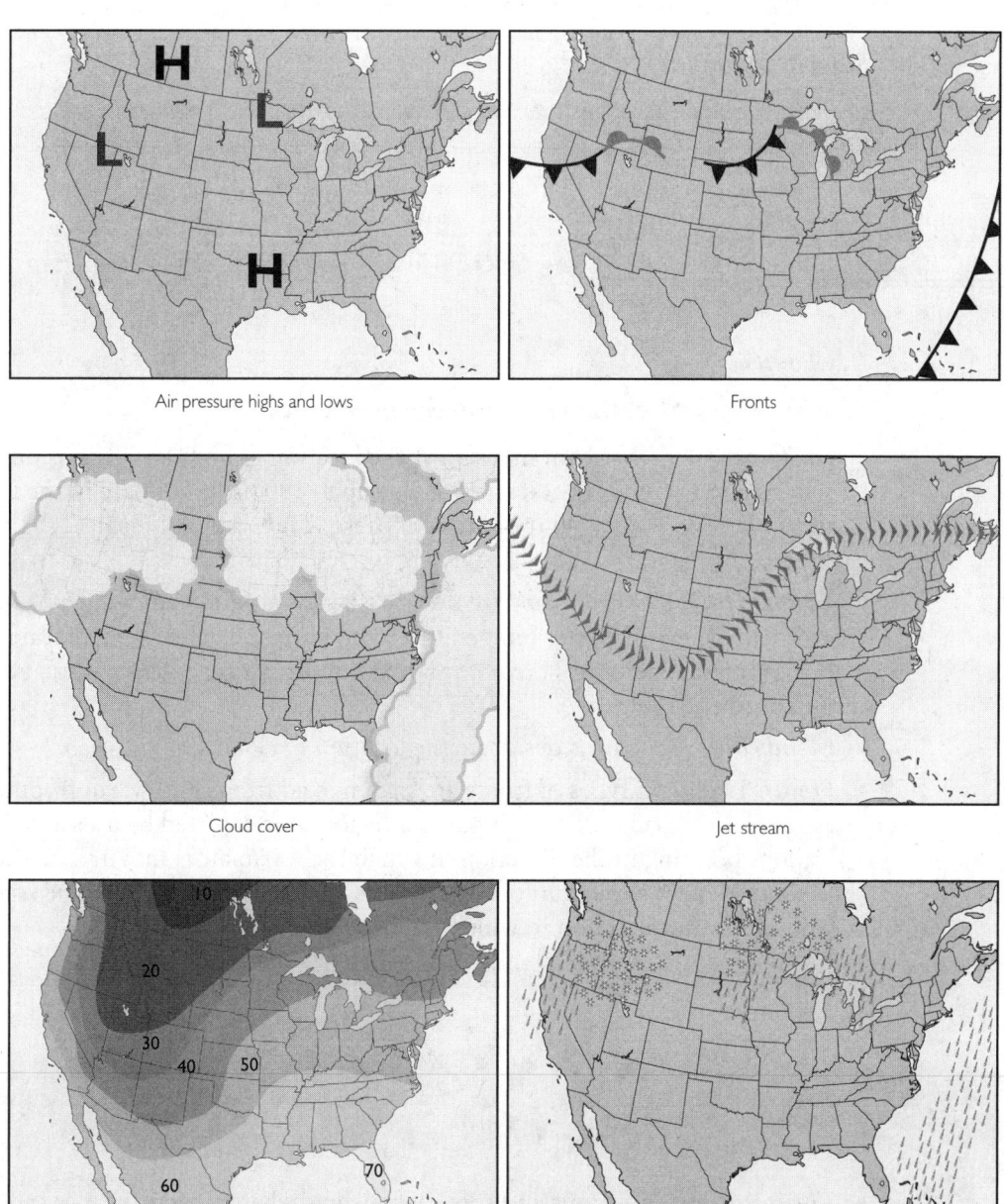

Air pressure highs and lows

Fronts

Cloud cover

Jet stream

Temperature highs in degrees Fahrenheit

Precipitation: rain and snow

Explain and Elaborate (15 minutes)

Discuss the Tools of Weather Prediction

→ Display the appropriate transparencies as needed.
→ Draw weather symbols on the board and label them.

Sample Questions

- What weather variables were on the weather maps you studied? (jet stream, temperature, cloud cover, precipitation, fronts, high and low pressure)
- What are some of the symbols found on the maps? What do you think they stand for?
- Why is it necessary for a meteorologist to use more than one map in order to predict the weather?

Key Points

Several factors affect the weather in North America:

- ***Jet stream.*** The jet stream map shows the location of high-level winds on the planet. These winds are at least 57 mi/h (up to 190 mi/h) and are in the upper atmosphere above 20,000 ft (4 mi up). These winds generally "steer" storms around the planet.
- ***Temperature.*** Bands of color are used to display variations in temperature, with reds being the hotter temperatures and blues the colder temperatures. Notice that temperatures are not random but form a very clear pattern across the continent.
- ***Cloud cover.*** Gray areas designate the location of clouds.
- ***Fronts.*** Two main types of fronts are shown, cold fronts and warm fronts. In a warm front, warm air is moving into a region. It is depicted by a series of red scallops pointing in the direction in which the warm air is moving. In a cold front, cold air is moving into a region. It is depicted by a series of blue triangles pointing in the direction in which the cold air is moving.

Cold front   Warm front

- ***Precipitation.*** Raindrops and snowflakes show where there is rain or snow.
- ***Pressure.*** The H's show the locations of large air masses with consistently high pressure. The L's show where large air masses with consistently low air pressure are located.

Discuss the General Relationships Among Weather Variables

→ Use the map transparencies as needed to show the general relationships between weather variables.

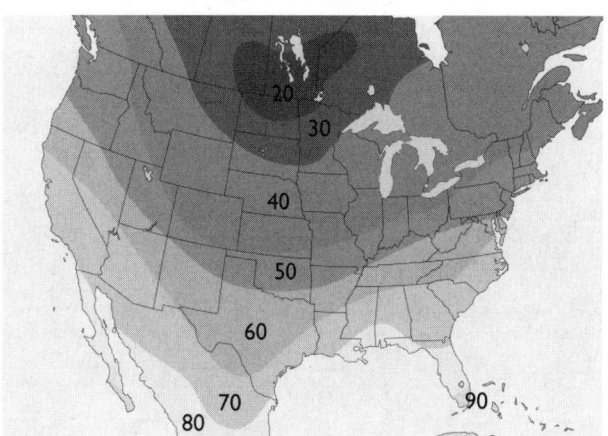

Sample Questions
- What connections did you find between the different maps when they were placed on top of each other?
- If you were going to try to predict whether it will rain in a certain city, which weather maps would you want to have? Explain.

Key Points

There are connections between the different maps:

- The precipitation map is most closely associated with low-pressure areas and with weather fronts of all kinds.
- The curves of the jet stream map generally match the curves of the temperature maps.
- The lows on the air pressure map are associated with the warm and cold fronts.
- Rain requires clouds overhead. However, the presence of clouds does not necessarily mean precipitation.
- The jet stream divides cold arctic air from warmer air in the south. As winter sets in, the jet stream moves lower down over the continent. Broadly speaking, you can predict the general location of the jet stream by examining the pattern created by the temperature bands and placing the jet stream between the warmer and colder temperatures.

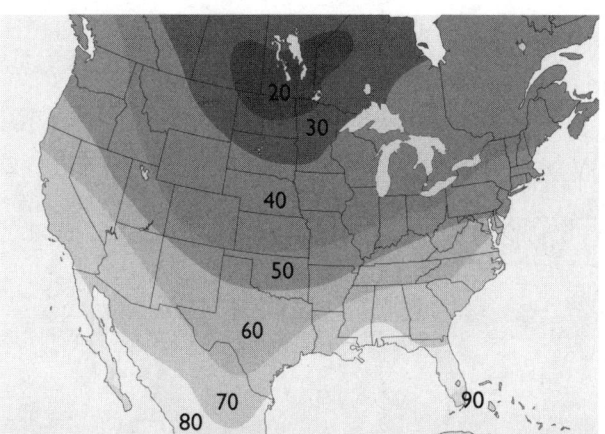

Temperature highs in degrees Fahrenheit

In the jet stream, air moves swiftly from west to east across the United States. Storms located in the atmosphere below the jet stream generally move from west to east across the continent. If you want to predict the weather in the United States, often it helps to look to the west.

Over the course of this unit, you will continue to explore what causes the weather and how you can predict weather from a variety of weather data.

Wrap-up

Key Question: What causes the weather?

- Weather consists of precipitation, clouds, winds, and temperature changes.
- Weather is the result of interaction among Earth, the atmosphere, water, and the Sun.

- The jet stream, temperature, cloud cover, weather fronts, precipitation, and air pressure are variables that are all tracked by meteorologists in order to forecast the weather.

Evaluate (5 minutes)

> **Check-in**
>
> What do weather maps keep track of? How do they help meteorologists?

Answer: Weather maps keep track of the jet stream, temperature, cloud cover, weather fronts, precipitation, air pressure, and other variables. Meteorologists can examine these maps and reach conclusions about what probably will happen weather-wise.

Homework

Assign the reading and exercises for Weather Lesson 1 in the student text.

LESSON 2 OVERVIEW

Raindrops Keep Falling
Measuring Liquids

Lesson Type
Lab:
  Pairs
Demo:
  Pairs

Key Ideas

This lesson focuses on proportional relationships, partly in preparation for gas laws, introduced later in the unit. Meteorologists measure rainfall by recording the height of rain that falls, in inches or millimeters. This is a valid reflection of rain amount because the volume of a container is proportional to its height. Height is a more useful measurement than volume for rainfall because it is not dependent on the overall size of the container used.

As a result of this lesson, students will be able to

- identify a proportional relationship
- describe several methods for solving a problem involving proportional variables
- explain why rainfall is measured in terms of height rather than volume

Focus on Understanding

- Many rulers do not start at zero. Students need to correct for the offset. If they do not, they will not obtain a straight line for the relationship between volume and height.
- This is a good opportunity to review accuracy, precision, and significant digits.
- For simplicity, we refer to directly proportional relationships in this unit simply as proportional relationships.

Key Terms

proportional
proportionality constant

What Takes Place

In this activity students explore the relationship between volume and height as they investigate the optimal way to measure rainfall. The lesson begins with a quick demonstration to simulate rain falling over different surface areas. During the activity, students measure the volume of rain in two containers to find specific rainfall amounts. They then plot volume versus height on a graph. The graph allows students to see the proportional relationship between the height and the volume of rain that falls into a container with parallel sides. Proportional relationships are discussed further in the next lesson.

Note: In the student text, a Math Spotlight focuses on proportional relationships.

Section I   Lesson 2 Raindrops Keep Falling   11

Materials

- student worksheet
- transparencies—ChemCatalyst, Volume Versus Height Graph

Demonstration materials (dry demo)

- commercial rain gauge (optional)
- several paper cutouts with different surface areas (e.g., binder paper, sticky note)

Demonstration materials (wet demo—optional)

- empty fish tank or other large transparent container
- two beakers of different size and a Florence flask that fit inside the container
- a foil or plastic tub that can sit on top of the container

Per pair

- 100 mL graduated cylinder or 600 mL graduated beaker (if available)
- 400 mL beaker, ungraduated (or cylindrical jar)
- 250 mL Florence flask
- wash bottle
- ruler marked in inches and centimeters (preferably with zero at the end)
- water, 500 mL

Setup

For the optional wet demonstration, you will need a large transparent container (e.g., a fish tank); a small beaker, a larger beaker, and a Florence flask to fit inside; and a foil roasting pan or plastic tub to place on top. Poke a lot of tiny, evenly spaced holes in the tub with a thumbtack (e.g., a 0.5 in.^{2} array). To watch it rain inside the large transparent container, place the plastic tub on top and fill it with water.

Graduated beakers, if available, may be used instead of graduated cylinders to speed up the volume measurements for the lab.

LESSON 2 GUIDE

Raindrops Keep Falling
Measuring Liquids

Engage (5 minutes)

Key Question: How do meteorologists keep track of the amount of rainfall?

> **ChemCatalyst**
>
> 1. How is rainfall usually measured? Describe the type of instrument you think is usually used.
> 2. Which of these containers would make the best rain gauge? Explain your reasoning.
>
> Bucket   Florence flask   Beaker

Sample Answers: 1. Rain is measured with a rain gauge, which looks a little like a graduated cylinder. 2. Each container shown has flaws. The narrow neck of the flask limits the ability of rain to enter the container. The bucket lacks graduations and it has sloped sides. The beaker has straight sides with somewhat imprecise graduations that measure volume, not height.

Discuss the ChemCatalyst

Sample Questions

- What units and instruments are used to track rainfall?
- What are some of the issues with the containers shown in the ChemCatalyst?
- Describe how you would measure rainfall. What kind of container would you use?
- How is volume different from height?

Explore (20 minutes)

Complete the Demonstration

➡ Simulate rainfall over a region, with either a wet or a dry demonstration. Show students a commercial rain gauge (optional).

Section 1   Lesson 2 Raindrops Keep Falling   13

Dry Demonstration

→ Show students several flat objects with different areas, such as a piece of binder paper, a circular coaster, and a sticky note. Place these objects on a table (or in a plastic tub). Ask students to imagine a rainstorm passing through the classroom with rain falling consistently on all these areas.

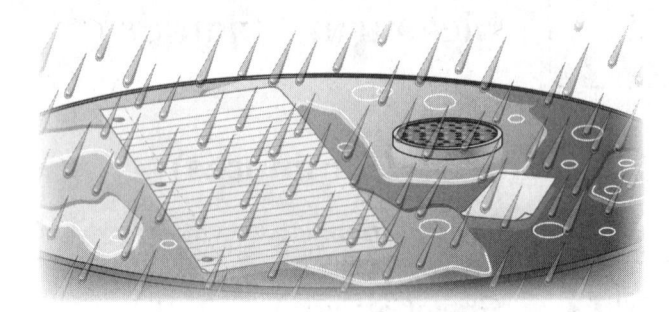

→ Ask students if the same amount of rain falls on each area. Ask them how they might measure the amount of rain that falls on the objects.

Wet Demonstration (optional)

→ Place a small beaker, a larger beaker, and a Florence flask in a large transparent container. Place the tub with holes on top of the container.

→ Ask students to predict the height of the water in the three small containers when you pour water into the tub.

→ Pour the water into the tub and let the students observe.

Introduce the Lab

→ Tell students they will measure different amounts of rain in one rain gauge and then repeat the procedure with a second rain gauge.

LESSON 2 LAB

Raindrops Keep Falling
Measuring Liquids

Name _____
Date _____ Period _____

Purpose
To experiment with rain gauges and to understand proportional mathematical relationships.

Materials
- 400 mL ungraduated beaker
- 250 mL Florence flask
- 100 mL graduated cylinder or 600 mL graduated beaker
- wash bottle
- ruler marked in inches and centimeters

Data Collection

1. Use the wash bottle to carefully pour water to a height of 2.0 cm into the 400 mL ungraduated beaker. Make sure you measure from the zero on the ruler.

2. Transfer the water to the graduated cylinder or graduated beaker. Measure and record the volume. Repeat this process for the heights of water given in the table.

Height (cm)	0 cm	2.0 cm	4.0 cm	6.0 cm	8.0 cm
Volume in beaker (mL)	0 mL	80 mL	165 mL	245 mL	330 mL
Volume in flask (mL)	0 mL	55 mL	95 mL	215 mL	235 mL

3. Repeat the procedure for the 250 mL Florence flask.

Graph of Data

1. On the graph below, plot the volume of rain in milliliters versus its height in centimeters for the 400 mL beaker. Draw a straight line through the points and label it "400 mL beaker."

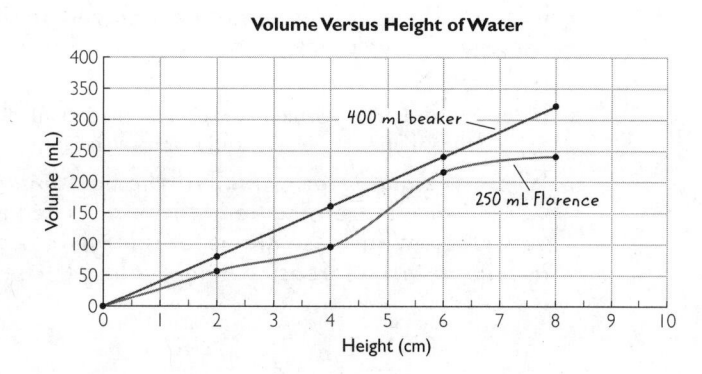

Volume Versus Height of Water

(Graph showing two lines labeled "400 mL beaker" and "250 mL Florence", with Volume (mL) on y-axis from 0 to 400 and Height (cm) on x-axis from 0 to 10.)

2. Explain why the data points lie roughly on a straight line.

Each time you increase the height by 1 cm, the volume changes by the same increment, ~40 mL.

3. What volume would you predict for 10.0 cm of water? Explain how you arrived at your answer. Use the data table and the graph to assist you in answering the question.

The volume would be ~400 mL. You get this by extending the line on the graph to 10 cm and reading the volume. You can also multiply 10.0 cm by 40 mL, because there are ~40 mL per centimeter. This gives 400 mL at 10.0 cm.

4. Plot the volume of rain in milliliters versus its height in centimeters for the 250 mL Florence flask on the same graph. Draw a best-fit curve through the points and label it "250 mL Florence flask."

5. Explain why the data points do not lie on a straight line for the Florence flask.

Each time you increase the height by 1 cm, the volume changes by a different amount depending on the diameter of the flask at that height.

6. **Analysis** Imagine that you use a 400 mL beaker as a rain gauge and your next-door neighbor uses a 600 mL beaker with a larger diameter. How should your results compare after an evening's rainfall? Explain your reasoning.

The height of the rain will be the same in both, but the volume will be larger for the 600 mL beaker because it has a larger diameter.

7. **Making Sense** These drawings show rain amounts at different times of the day for three different sizes of rain gauge. Explain the variations in the height and volume of the rainfall.

12 P.M.   6 P.M.   12 P.M.   3 P.M.   6 P.M.   12 P.M.

The height of the rain is the same at the same time of day because the height does not depend on the size of the rain gauge. However, the volumes of rain vary with the diameter of the rain gauge.

8. **If You Finish Early** Gather more data for the 250 mL Florence flask. Plot the data. Explain the variation in volume with increasing height.

Where the diameter is small at the bottom and top, the change in volume with increasing height is small. The line on the graph is not very steep. Where the diameter is large in the middle, the change in volume with increasing height is large. The line on the graph is steeper.

Explain and Elaborate (15 minutes)

Discuss the Graph

[T] ⇒ Display the transparency Volume Versus Height Graph. At the appropriate time in the discussion, draw a line or curve through each set of points.

Sample Questions

- Why do the data points lie more or less along a straight line for the beaker but not for the Florence flask?
- Which would make a better rain gauge, the beaker or the Florence flask? Explain your thinking.
- Explain how you made your predictions for the volume of 10.0 cm of water in the 400 mL beaker.

Key Points

The data points show that volume increases in a steady and predictable way in relation to the height of the beaker. There are several good ways to make volume predictions for the beaker from the data. They are outlined here:

- Use the data in the data table to estimate the volume at a different height.
- Use the straight line on the graph to predict other data points.
- Calculate new volume values by using proportions. For example, in the same beaker, 50 cm of rain will produce 10 times as much volume as 5 cm of rain.

All these methods are possible because the height and volume of a container are proportional to each other for the container. The data points for the beaker lie on a straight line going through zero. Whenever a graph of two variables results in a straight line that passes through the origin, (0, 0), the two variables are proportional to each other. The data points for the Florence flask do not lie on a straight line. Thus, volume is not proportional to height for the flask.

> **Proportional:** Two variables are directly proportional when you can multiply the value of one by a constant to obtain the value of the other.

The graphed line and the math associated with the line represent an ideal. This is why the data points measured in the experiment for the beaker aren't all *exactly* on the line drawn. There is always some error in measuring data. The line describes what would happen if your data points were the result of perfect measurements. Math and measurement do not always match exactly. Nevertheless, the math allows you to make accurate predictions.

Explore Proportional Relationships

Sample Questions

- How are the volume and height of a container with parallel walls related to each other mathematically?
- If the amount of rainfall increases, do both the volume and the height of water in the rain gauge keep track of this increase? Explain your thinking.
- Why do meteorologists report the height of rainfall instead of the volume of rainfall?
- Can you think of another proportional relationship?

Key Points

The volume of rainfall increases regularly in relation to the height of the rainfall. This is because height is proportional to volume.

The *height* of rain collected in a rain gauge does not depend on the diameter of the container. A rainstorm that drops 1 in. of rain in one container drops 1 in. of rain in any other container (provided the walls of the containers are parallel).

The *volume* of rain collected does depend on the diameter of the container. After a storm, a container with a larger diameter will have more water in it than one with a smaller diameter.

If meteorologists reported rainfall in terms of volume, they would all need to use a rain gauge with the exact same size and shape in order to be able to compare measurements. Thus, meteorologists report rainfall in inches, centimeters, or millimeters.

There are countless examples of proportional relationships in the world around you. For example:

- The number of steps in a staircase is proportional to the total height of the staircase.
- The distance traveled by an automobile going a certain speed is proportional to the amount of time it is driven.
- The number of pages in a book is proportional to the thickness of the book.
- The number of eggs you have is proportional to the number of cartons of eggs you have.
- The relationship between centimeters and inches is proportional.

All these proportional relationships can be expressed mathematically.

- The relationship between volume and height of rain gauges can be expressed mathematically by this formula:

    Volume = (area of base) · (height)

- The relationship between distance and time traveled by an object can be expressed by this formula:

    Distance = (speed) · (time)

- The relationship between centimeters and inches is expressed as

    Inches = (2.5) · (centimeters)

In each case, one number in the equation does not change. If the area of the base does not change, that becomes a constant in the first equation. Similarly, if the speed is steady (e.g., 55 mi/h on the freeway), speed becomes the constant in the second equation. This unchanging number is often referred to as the proportionality constant. You can see that if the variable on one side of the equation is doubled, or tripled, or quadrupled, the variable on the other side of the equation must also be doubled, tripled, or quadrupled in order to satisfy the equality.

> **Proportionality constant:** The number that relates two variables that are proportional to each other. It is represented by a lowercase k.

Wrap-up

Key Question: How do meteorologists keep track of the amount of rainfall?

- Meteorologists measure the height of rain, because the volume of rain is directly proportional to the height of the water measured.
- Graphs of two variables that are proportional are always a straight line through the origin.
- When one variable is proportional to another, it is possible to make accurate predictions of other values when one data point is known.

Evaluate (5 minutes)

> **Check-in**
>
> Suppose you find that a cylindrical rain gauge contains a volume of 8 mL of rain for a height of 2 cm of rain. Describe how you might figure out the volume of rain for a height of 10 cm of rain in this same container.

Answer: Graph the data point and draw a line through this point and the origin, (0, 0). Then find the volume for 10 cm. Or use proportionality. Because the new height, 10 cm, is 5 times 2 cm, the new volume will be 5 times 8 mL, or 40 mL.

Homework

Assign the reading and exercises for Weather Lesson 2 in the student text.

LESSON 3 OVERVIEW

Having a Meltdown
Density of Liquids and Solids

Lesson Type
Lab:
 Groups of 4

Key Ideas

When a substance changes phase, its density also changes. For instance, water has different densities depending on whether it is in the form of snow, ice, or rain. Density relates mass and volume. If you know the density of snow, you can calculate the volume of water obtained when a given volume of snow melts.

As a result of this lesson, students will be able to

- make density calculations, converting volumes of liquids and solids
- explain how phase changes affect the density of a substance
- use density equations to calculate the volume of water in a sample of snow or ice

Focus on Understanding

- For the relationship between mass and volume, the proportionality constant, k, is also the density, D. The proportionality constant does not always have a name.
- Many students have trouble rearranging equations to solve for the value they need.

What Takes Place

SCiLINKS NSTA
Topic: Density
Visit: www.SciLinks.org
Web code: KEY-105

Students begin class by exploring the idea that snow and liquid water have different densities. They then design a procedure to measure the volume and mass of different amounts of water in order to determine the density of water. Finally, they use their data to calculate the volume of water obtained when a given volume of snow melts. Students practice working with proportional relationships. The worksheet gives them the opportunity to use the formula $D = \frac{m}{V}$, which they learned in Unit 1: Alchemy. This time they also see the relationship graphically, with a graph relating mass to volume. They can use a Math Card, which summarizes calculations and conversions for proportional relationships. The fact that the slope of the line is the proportionality constant is covered in the reading for this lesson in the student text.

Materials

- student worksheet
- Math Cards—Density, and Triangle Instructions; one each per student (optional)
- transparency—Phases of Water Graph

Per group of 4

- 25 mL graduated cylinder
- scale
- wash bottle of water

20 *Living By Chemistry Teacher Guide*   Unit 3 Weather

Setup

If snow is available where you live, this class could easily be modified to allow students to collect, measure, and melt snow and calculate the densities of both water and snow.

The Math Cards are an optional tool. They give students a scaffold until they become more familiar with the proportional relationships in this unit. All the Math Cards for this unit have been included in the *Teaching and Classroom Masters* pages for this lesson. Only the first one pertains to this lesson, but you might choose to copy and cut them all at the same time, saving the rest for future lessons. Students will also need the Triangle Instructions card, which describes how to use the Math Cards.

LESSON 3 GUIDE

Having a Meltdown
Density of Liquids and Solids

Engage (5 minutes)

Key Question: How much water is present in equal volumes of snow and rain?

> **ChemCatalyst**
>
> Water resource engineers measure the depth of the snowpack in the mountains during the winter months to predict the amount of water that will fill the lakes and reservoirs the following spring.
>
> 1. Do you think 3 inches of snow is the same as 3 inches of rain? Explain your reasoning.
> 2. How could you figure out the volume of water that will be produced by a particular depth of snow?

Sample Answers: 1. Students might say that 3 inches of snow doesn't contain as much water as 3 inches of rain because the snow is fluffier or less dense than rain. 2. Students might suggest melting the snow to measure the volume of water contained in it.

Discuss the ChemCatalyst

➡ Discuss how to compare amounts of snow and rain.

Sample Questions

- How is snow different from rain? How is it the same?
- Does 10 g of snow have the same mass as 10 g of water? Explain.
- Which do you think will be denser: snow, ice, or rain? Explain your reasoning.

Explore (15 minutes)

Introduce the Lab

➡ Tell students that they will measure the mass of four different volumes of water and use their data to determine the density of water.

➡ Leave plenty of time for the Explain and Elaborate portion of the class.

LESSON 3 LAB

Having a Meltdown
Density of Liquids and Solids

Name _____
Date _____ Period _____

Purpose
To compare the amount of water present in volumes of snow and rain by using their density values.

Materials
- 25 mL graduated cylinder
- scale
- wash bottle

Data Collection
Determine the mass of four different volumes of water. The volumes can be any amount between 1 mL and 25 mL. Record your measurements in the table.

Mass of the empty graduated cylinder: _____

Measured			Calculated
Mass of water plus graduated cylinder (g)	Mass of water (g)	Volume of water (mL)	Density = $\frac{mass}{volume}$ (g/mL)
varies	5.5 g	5.5 mL	1.0 g/mL
varies	10.2 g	10.2 mL	1.0 g/mL
varies	15.7 g	15.7 mL	1.0 g/mL
varies	25.0 g	25.0 mL	1.0 g/mL

Note: These data are idealized. There will be more variation in real data due to measurement error, but values in the second and third columns should be very close.

Questions

1. What is the volume of 12 g of rain? How do you know?

   12 g of rain occupies 12 mL, because there is 1 g per milliliter.

2. The mass and volume data in these tables were collected for snow and ice. Use the data to figure out the density of snow and ice for these measurements.

Snow

Mass	Volume	Density
1.7 g	3.4 mL	0.5 g/mL
3.7 g	7.4 mL	0.5 g/mL
7.9 g	15.8 mL	0.5 g/mL
10.2 g	20.4 mL	0.5 g/mL

Ice

Mass	Volume	Density
2.2 g	2.4 mL	0.92 g/mL
5.9 g	6.4 mL	0.92 g/mL
9.2 g	10.0 mL	0.92 g/mL
20.2 g	22.0 mL	0.92 g/mL

3. Plot the mass and volume data for rain, snow, and ice on the same graph. Draw the best straight line through each set of data points. Label them "rain," "snow," and "ice."

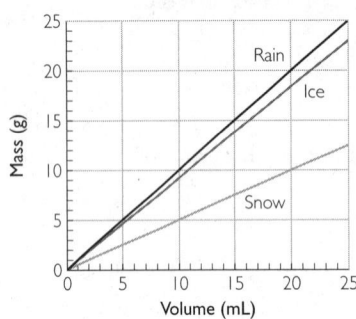

Mass Versus Volume of Water

[Graph showing Mass (g) vs Volume (mL) with three lines labeled Rain, Ice, and Snow, where Rain has the steepest slope, Ice is slightly less steep, and Snow has the least steep slope.]

4. What is the volume of 12 g of snow? How do you know? What is the volume of 12 g of ice? Show your work.

   12 g of snow occupies 24 mL, because there is 0.5 g per milliliter. You can also read this off the graph.

   $$\text{Use } D = \frac{m}{V} = \frac{12 \text{ g}}{0.92 \text{ g/mL}} = 13 \text{ g}$$

5. **Compare.** Place 12 g of rain, 12 g of snow, and 12 g of ice in order of increasing volume. Use density to explain the order.

   In order of increasing volume, rain < ice < snow. Rain is densest. Thus, the volume of 12 g of rain is the smallest. Snow is the least dense. Thus, the volume of 12 g of snow is the largest.

6. **Compare.** Examine the lines on the graph for rain, snow, and ice. How is the steepness of each line related to the density of each substance?

   The steeper the line, the denser the substance.

7. **Calculate.** Imagine a 24 mL sample of snow from the mountains. When the snow melts, what is the volume of liquid water? Assume that the snow has density 0.5 g/mL.

   There is 0.5 g of snow per milliliter of snow. Thus, 24 mL of snow is 12 g of snow. When 12 g of snow melts, you have 12 g of liquid water. There is 1.0 g of water per milliliter of water. Thus, 12 g of water is 12 mL of water. The volume of water is half the volume of snow.

8. **Making Sense** Explain how you can use the graph to compare the mass of water in equal volumes of snow and rain.

   For a specified volume, draw a vertical line. When you reach the slanted line for water, draw a horizontal line to the y-axis to find the mass. Do the same for snow. The mass for water will be larger because water is denser.

9. **If You Finish Early** If you have 10 cm of snow with volume 40 mL and density 0.5 g/mL, how many inches of rain is this?

   The density of rain is twice the density of snow, so the height will be half as much: 5 cm of rain, or about 2 in.

Living By Chemistry Teaching and Classroom Masters: Units 1–3
© 2010 Key Curriculum Press

Unit 3 Weather 209
Lesson 3 • Worksheet

Explain and Elaborate (20 minutes)

Discuss the Density Changes Associated With Phase Changes

→ Display the transparency of the graph showing mass versus volume for water in different phases.

Sample Questions

- How did you figure out the density of liquid water? What value did you obtain?
- How do the densities of different volumes of liquid water compare?
- How do the densities of snow and ice compare to the density of liquid water?
- What happens to the density of a substance when it changes phase? What evidence do you have to support this idea?
- How do the graphs differ for ice, water, and snow? How are they the same? (They have different slopes, or different steepness; each is a straight line going through the origin.)

Key Points

You can determine the density of water by measuring the mass of a certain volume of water. The density, D, is the mass, m, divided by the volume, V:

$$D = \frac{m}{V}$$

The ratio of mass to volume for liquid water is 1.0 g/mL (at room temperature and sea level). The value is the same for any amount of water. The density of a small glass of water and a large pool of water is the same, 1.0 g/mL as long as both are the same temperature. In both cases, 1 mL of water will have a mass of 1 g.

The densities of snow and ice are less than the density of water. Ice has a density of 0.92 g/mL. In other words, 1 mL of ice has a mass of only 0.92 g, so it floats in liquid water. Snow is even less dense, approximately 0.50 g/mL depending on the type of snow.

The graph you created allows you to compare the densities of the different phases of water. The differences in density for the three phases are reflected in the steepness of the lines. The steeper the line made by connecting the data points, the greater the density of the substance.

Discuss the Proportional Relationship $m = DV$

Sample Questions

- If you know the volume of a sample of snow, how can you find its mass without using a scale?
- What evidence do you have that mass and volume are proportional to each other?
- How are density units related to the units for mass and volume?

Key Points

The relationship $D = \frac{m}{V}$ can also be written as $m = DV$. When density is constant, the mass and volume are proportional. The proportionality constant is the density. Notice that the unit of mass is grams, the unit of volume is milliliters or cubic centimeters, and the unit of density is g/mL or g/cm^3.

For any proportional relationship, the graph is a straight line that passes through the origin, (0, 0). The three lines students graphed on the worksheet represent proportional relationships. The slope of each line is equal to the proportionality constant.

Complete Worked Examples

➡ Write the following sample problems on the board and have students solve them before discussing the solutions. You can also revisit the ChemCatalyst questions at this time.

➡ Display the transparency of the graph for student reference.

➡ Hand out a Density Math Card and a Triangle Instructions Card to each student (optional). Explain that these cards are a guide to using the density equation to solve problems for any of the three variables. Show students how to use the cards.

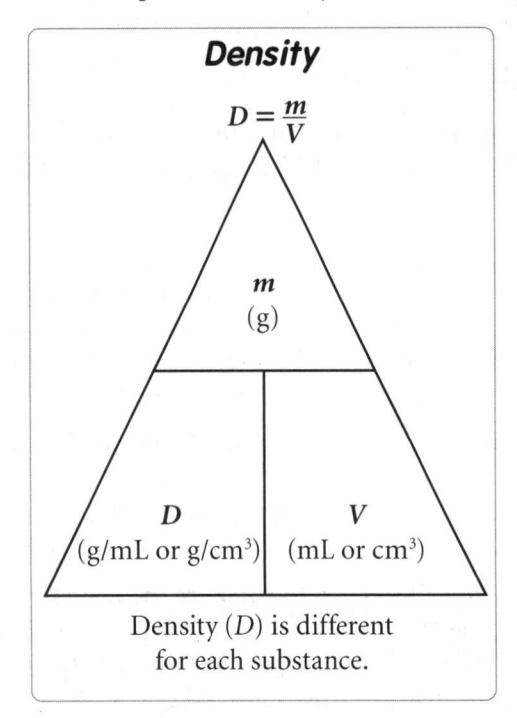

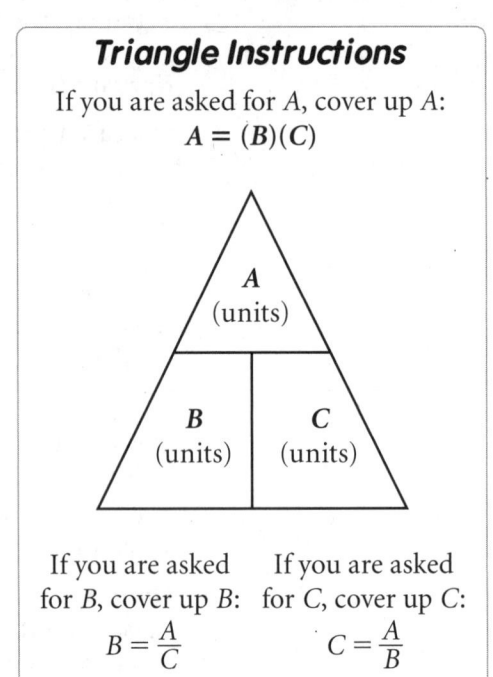

Density

$D = \dfrac{m}{V}$

- m (g)
- D (g/mL or g/cm^3)
- V (mL or cm^3)

Density (D) is different for each substance.

Triangle Instructions

If you are asked for A, cover up A:

$A = (B)(C)$

- A (units)
- B (units)
- C (units)

If you are asked for B, cover up B:

$B = \dfrac{A}{C}$

If you are asked for C, cover up C:

$C = \dfrac{A}{B}$

Sample Questions

- If you have the same mass of rain and snow, will the volumes occupied by the rain and the snow be the same or different? Explain.
- If you have 100 mL of snow, how would you determine the volume of the same mass of rain?

Key Point

Scientists measure snowpack in terms of depth—meters or feet—and then make conversions to obtain the volume of water. This is another application of the law of conservation of mass.

Example 1

Imagine that you have a box with volume 14.5 mL. What mass of ice will just fill this box?

Solution

The mass is equal to the density times the volume.

$$m = DV$$

$$m = 0.92 \text{ g/mL} \cdot 14.5 \text{ mL} = 13 \text{ g of ice}$$

Example 2

You have 12 g of snow with density 0.50 g/mL. What volume does this snow occupy in milliliters?

Solution

Solve the mathematical equation for volume, then substitute known values.

$$m = DV$$

$$V = \frac{m}{D}$$

$$V = \frac{12 \text{ g}}{0.50 \text{ g/mL}} = 24 \text{ mL}$$

Example 3

If you have 100 mL of snow, what volume of water do you have?

Solution

When the snow melts, the volume will change but the mass will stay the same. First, determine the mass of snow. Assume the snow has a density of 0.5 g/mL.

$$m = DV = 0.5 \text{ g/mL} \cdot 100 \text{ mL} = 50 \text{ g of snow}$$

When the snow melts, you get the same mass of water.

50 g of snow = 50 g of water

Finally, determine the volume of liquid water. The density of water is 1.0 g/mL.

$$V = \frac{m}{D} = \frac{50 \text{ g}}{1.0 \text{ g/mL}} = 50 \text{ mL}$$

Because water is denser, the volume of liquid water is less than the volume of snow.

Wrap-up

Key Question: How much water is present in equal volumes of snow and rain?

- In density calculations, mass and volume are proportional to each other.
- The formula $D = \frac{m}{V}$ can also be written $m = DV$.

- When a substance changes phase (from solid to liquid to gas), its density changes. The mass stays the same, and the volume changes.
- Water is denser than snow. Ice is less dense than water.

Evaluate (5 minutes)

> **Check-in**
>
> 1. Imagine that you have equal masses of snow and rain. Which has a greater volume? Explain your thinking.
> 2. What is the mass of 14 mL of rainwater?

Answers: 1. Snow is less dense than rain and will occupy a larger volume for a given mass. 2. The mass of 14 mL of rainwater is 14 g.

Homework

Assign the reading and exercises for Weather Lesson 3 in the student text.

LESSON 4
OVERVIEW

Hot Enough
Thermometers

Lesson Type
Lab:
Groups of 4

Key Ideas

The volume of matter changes in response to changes in temperature. Almost all substances expand on heating and contract on cooling. A thermometer is constructed to measure the volume change of a substance in response to temperature changes. Then it is calibrated to reflect temperature. Physical phenomena that are reproducible, such as the boiling point and freezing point of water, are used to construct temperature scales.

As a result of this lesson, students will be able to

- create a thermometer and a temperature scale
- describe how a thermometer works
- explain the Fahrenheit and Celsius temperature scales

Focus on Understanding

- The idea that thermometers do not measure temperature directly but instead measure the expansion of a liquid or a gas may not have occurred to students.
- Students may have trouble grasping the idea that a temperature scale is fairly arbitrary and that they can create any temperature scale they want provided it accurately reflects at least two standard repeatable temperatures.
- The gas thermometer can be difficult to understand, because the trapped gas is not visible.

SciLinks NSTA
Topic: Temperature Scales
Visit: www.SciLinks.org
Web code: KEY-304

Key Terms

melting point (melting temperature)
boiling point (boiling temperature)
degree

What Takes Place

Students use a rudimentary thermometer made from glycol, a plastic straw, a vial, and a rubber stopper. This instrument is placed in water of varying temperatures, and the height of the liquid is marked. Students calibrate their thermometer and figure out the current temperature in the room. They also make a rudimentary gas thermometer and explore the expansion and contraction of air in response to temperature. Students are introduced to the Fahrenheit and Celsius temperature scales.

Section I   Lesson 4 Hot Enough

Materials

- student worksheet
- cork borer

Per group of 4

- small glass vial (~10–20 mL capacity for glycol thermometer)
- 250 mL beakers (3) (for boiling water, ice water, ice water with salt)
- hot plate
- ice (to fill two beakers)
- salt, 1 tbsp (for ice bath)
- small metric ruler
- ethylene glycol (antifreeze), ~25 mL per thermometer
- 1-hole rubber stopper or rubber septum to fit over the vial
- rigid, clear, hard plastic straw to go through septum or fit into the hole in the stopper, ~15 cm long and 0.5 cm diameter
- small amount of petroleum jelly to lubricate the stopper and the straw, if using stopper
- test tube holder, wire
- fine-point permanent marker to mark the straw
- alcohol (to remove markings from the straw)
- 10 mL graduated cylinder (for gas thermometer)
- several drops of red, green, or blue food coloring

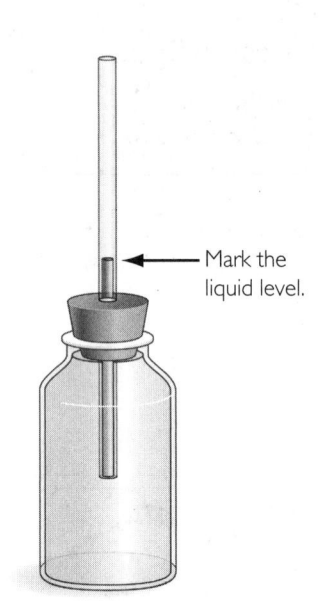

Mark the liquid level.

Setup

You will need about 2 lb of crushed ice per class. Assemble eight ethylene glycol thermometers. Alternatively, you can assemble one thermometer as an example and have groups make their own. Use a cork borer to put a hole through each rubber septum or stopper. Put the hard plastic straw through the hole. (You may need petroleum jelly to help the straw slip through. Take care not to get petroleum jelly inside the straw.) Fill the vials with ethylene glycol so that the liquid level is all the way to the top. You might want to add a few drops of food coloring to the ethylene glycol so it is easier to see in the straw. Put the septum with the straw on the vial so that liquid rises in the straw.

You can have students place 250 mL beakers with water on hot plates before the ChemCatalyst so they do not waste time waiting for the water to boil.

Cleanup

You can use the previously assembled thermometers for another class. Use alcohol to wipe off any marks left on the straws.

LESSON 4 GUIDE

Hot Enough
Thermometers

Engage (5 minutes)

Key Question: How is temperature measured?

ChemCatalyst

The weather forecast in Moscow, Russia, calls for a 60% chance of precipitation with highs reaching 30 °C, while in Washington, D.C., the weather forecast calls for a 70% chance of precipitation with highs reaching 50 °F.

1. Which city will be warmer? Explain your thinking.
2. Do you think it will rain or snow in either of the two cities? Explain your reasoning.

Sample Answers: 1. Students might confuse the two temperature scales. Moscow will be warmer because 30 °C (equivalent to 86 °F) is warmer than 50 °F. 2. Some students will confuse the two scales and say it will snow in Moscow. However, it won't snow in either place because it is too warm.

Discuss the ChemCatalyst

➡ Solicit students' ideas about how a thermometer works.

Sample Questions

- What weather did you predict for the two cities? Why?
- What is the difference between degrees Celsius and degrees Fahrenheit?
- What is temperature? How is it measured?
- Why is a liquid used in a thermometer? (Actually, students may be more familiar with digital thermometers, which rely on the fact that electron resistance changes with temperature.)

Explore (20 minutes)
Introduce the Activity

➡ Briefly go over the procedure for the liquid thermometer. Show students the pre-assembled ethylene glycol thermometer.

➡ Show students how to hold the thermometer with a test tube holder in water at different temperatures. The thermometer should not touch the sides or bottom of the beaker. Note that the bottom of the beaker on the hot plate can get hotter than boiling water and that prolonged contact with it can damage the thermometer.

➡ Briefly go over the procedure for the gas thermometer. Show students how to measure the gas volume using the markings on the graduated cylinder.

➡ Have students work in groups of four.

LESSON 4 LAB
Hot Enough
Thermometers

Name _____
Date _____ Period _____

Purpose
To examine how the volumes of a liquid and a gas change in response to temperature.

Part 1: Liquid Thermometer
Materials

- small glass vial
- 250 mL beakers (3)
- hot plate
- ice
- salt, 1 tbsp
- small metric ruler

- ethylene glycol (antifreeze), ~25 mL per thermometer
- rubber septum or rubber stopper
- clear plastic straw
- test tube holder, wire
- fine-point permanent marker
- alcohol (to remove markings)

Procedure
Use the permanent marker to mark the level the liquid reaches in the straw in the ethylene glycol thermometer for these five conditions:

- room temperature
- vial warmed by your hand
- ice water
- 200 mL ice water with 1 tbsp salt
- boiling water (thermometer should not touch the bottom of the beaker)

← Mark the liquid level.

Observations and Analysis

1. What did you generally observe when you warmed and cooled the thermometer?

   Liquid moves down the straw when it is cooled and up the straw when it is warmed.

2. What is happening to the liquid in the vial to make it move up and down in the straw?

   The volume of the liquid changes in response to changes in temperature.

3. Create a scale for the thermometer.

   a. Assign numbers for the places you marked on the straw for boiling water and ice water. What numbers did you choose and why?

   Sample answer: 0 for ice water and 100 for boiling water

   b. Based on your newly created temperature scale, estimate the temperature in the room. How did you arrive at your answer?

   Sample answer: 25°, because it is about one-quarter of the way up from the mark for ice water to the mark for boiling water

Living By Chemistry Teaching and Classroom Masters: Units 1–3
© 2010 Key Curriculum Press

Unit 3 Weather 213
Lesson 4 • Worksheet

32 *Living By Chemistry Teacher Guide* Unit 3 Weather

4. The Fahrenheit scale and the Celsius scale are shown here side by side:

   a. What is the temperature of the room in degrees
      Celsius? __20 °C__ Fahrenheit? __68 °F__

   b. What is body temperature in degrees
      Celsius? __37 °C__ Fahrenheit? __98.6 °F__

   c. Which is hotter, 30 °C or 30 °F? Explain your reasoning.
      30 °C is hotter. Celsius units are larger than Fahrenheit units.

   d. Estimate what 50 °C would be on the Fahrenheit scale. *~122 °F*

   e. The formula for conversion from degrees Celsius to degrees Fahrenheit is $F = \frac{9}{5}(C) + 32$. Check your answer to part d by performing the calculation.

      $F = \frac{9}{5}(50\ C) + 32 = 122\ °F$

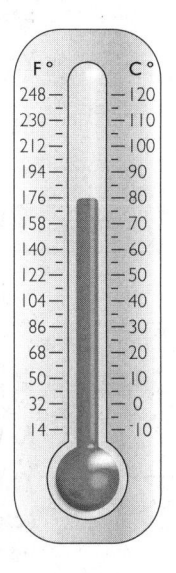

Part 2: Gas Thermometer

Materials

- 250 mL beakers (3)
- ice
- test tube holder, wire
- hot plate
- 10 mL graduated cylinder
- food coloring

Procedure

1. Put 200 mL of water with one or two drops of food coloring into a 250 mL beaker. Heat the water to about 80 °C. Place room temperature water in a second beaker. Place crushed ice into a third beaker.

2. Hold the 10 mL graduated cylinder with a test tube clamp and invert it in the hot water for at least a minute. Make sure the mouth of the cylinder is almost touching the bottom of the beaker.

3. Quickly move the graduated cylinder to the room temperature water. Make sure the mouth of the cylinder is almost touching the bottom.

4. Keep the graduated cylinder in the second beaker as you add ice to the water. Record your observations for all three situations.

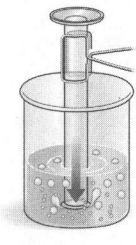

Analysis

1. Explain how you can use the air sample trapped inside the graduated cylinder as a thermometer.

   The air trapped inside expands when it is heated and contracts when it is cooled, just like the liquid in a thermometer.

2. **Making Sense** Describe how a thermometer works.

   A thermometer relies on changes in the volume of a liquid (or gas) at various temperatures. These changes are marked, and a regular scale is created.

Explain and Elaborate (15 minutes)
Discuss the Heating and Cooling of Matter

➡ You might want to sketch the illustrations below for the gas thermometer discussion.

Sample Questions
- How does a thermometer work?
- Explain the construction of the homemade liquid thermometer. Why is a straw needed?
- Provide evidence that air is trapped inside the graduated cylinder when it is held upside down in the water.
- Explain how you can use the sample of trapped air to construct a thermometer.

Key Points

The volume of matter changes in response to changes in temperature. Matter generally expands when heated and contracts when cooled. This property can be used to measure temperature, because the volume of liquid in a thermometer changes as the temperature changes.

Many thermometers contain a liquid, such as alcohol or mercury. The thermometer has a reservoir of liquid connected to a thin glass tube. The glass tube is needed because the volume change is relatively small. Narrowing the area makes the height change easier to detect.

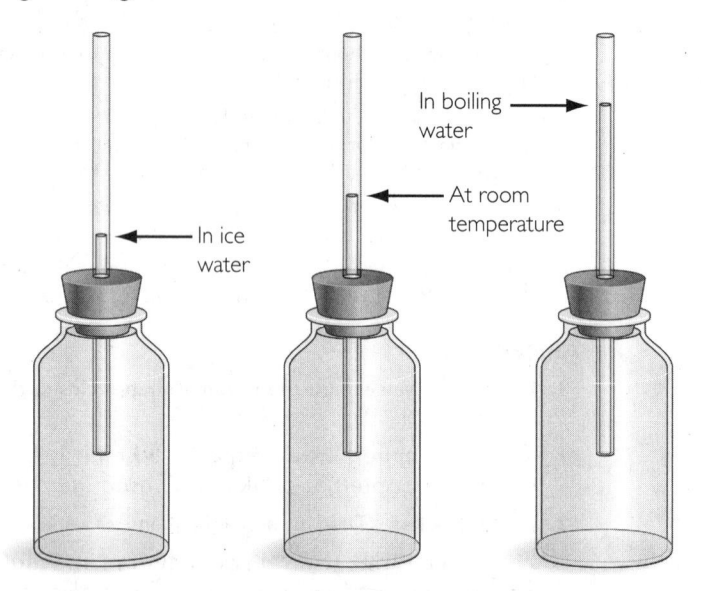

Gases expand and contract as the temperature changes. Water does not fill the inverted graduated cylinder when you push it into the water, evidence that air is trapped inside. As the sample of trapped air is warmed in the hot water, air bubbles escape. The air trapped inside expands to fill the entire volume. When the graduated cylinder with hot air is placed in water at room temperature, the air sample cools and water rises in the graduated cylinder. This is evidence that the air sample contracts as it cools, leaving room for some water to move up into the graduated cylinder. The volume occupied by the air decreases even more when the air is cooled further with ice.

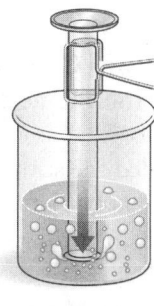

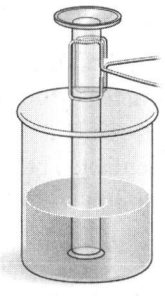

In hot water   In room-temperature water   In ice water

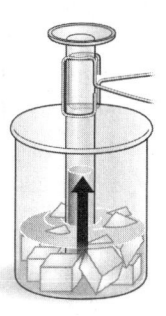

Discuss How Temperature Is Measured

Sample Questions

- If you know the volume or height of a liquid at two different temperatures, can you predict the temperature for a third volume or height? Explain your thinking.
- How can you use a trapped gas to measure temperature?
- How reliable is boiling water in setting a temperature scale? What about ice water?

Key Points

To set a temperature scale, you need at least two measurements. It is best to choose temperatures that are reproducible to set the scale. For example, ice water and boiling water are at very specific temperatures at sea level (the temperatures vary at different altitudes). In contrast, body temperature and room temperature are not good choices because both can vary from one moment (or one person) to the next.

> **Melting point or melting temperature:** The temperature at which a substance melts or freezes. At this temperature, both solid and liquid phases of the substance are present.
>
> **Boiling point or boiling temperature:** The temperature at which a substance boils or condenses. At this temperature, both liquid and gas phases of the substance are present.

Note: These definitions of melting point and boiling point refer to phases at equilibrium. Equilibrium is covered in Unit 6: Showtime.

Once you have set a scale with two points, you can determine other temperatures. Imagine that the temperature of ice water is called 0° and that of boiling water is called 60°. You can then make a mark halfway between these points and call it 30°. You can continue to divide up the scale and label it. On this particular scale, room temperature would be about 12°. Once two temperature points are noted on a scale, it is possible to figure out where any other temperature would be on that scale.

Relate the History of the Thermometer

➡ Share all or part of the following information with your class. Students can name their own temperature scales, just as Fahrenheit and Celsius did.

Key Points

In 1724, German physicist Daniel G. Fahrenheit invented the first modern thermometer—the mercury thermometer. To set his scale, he called the temperature of an ice/salt mixture 0° and called his own body temperature 96° Fahrenheit. Then he divided the scale into single degrees. (The scale has since been recalibrated so that normal body temperature is now 98.6 °F.) On his scale, the freezing point of pure water happens to occur at 32° (and the boiling point at 212°).

> **Degree:** The increment of temperature that corresponds to one unit on a thermometer. The size of a degree depends on the temperature scale used.

In 1747, Anders Celsius, a Swedish astronomer, created a thermometer with a different scale. Celsius used 0° and 100° for the melting point of snow and the boiling point of water, respectively. The Celsius temperature scale is now part of the metric system of measurement (SI). Most of the world and most scientists measure temperature in degrees Celsius, °C. It was formerly called the centigrade scale.

Discuss the Relationship Between °C and °F

Sample Questions

- Which units are larger, Celsius degrees or Fahrenheit degrees? (Celsius degrees)
- Would you be colder at 0 °F or 0 °C? (0 °F)
- When a thermometer reads 100 °C, what is the temperature in degrees Fahrenheit? (212 °F)
- Is there any advantage to one scale or the other?

Key Point

Degrees Celsius and degrees Fahrenheit both measure the same thing (temperature). Celsius units are larger than Fahrenheit units. Thus, a change of one degree Celsius represents a greater change in temperature than does a change of one degree Fahrenheit. To convert from degrees Celsius to degrees Fahrenheit, use this formula:

$$°F = \frac{9}{5}(°C) + 32 \quad \text{or} \quad °F = 1.8(°C) + 32$$

Wrap-up

Key Question: How is temperature measured?

- The volume of matter changes in response to changes in temperature.
- Almost all substances expand on heating and contract on cooling.
- The change in volume of a liquid can be used to measure temperature changes.
- The relationship between the Fahrenheit scale and the Celsius scale is described by the formula $°F = \frac{9}{5}(°C) + 32$.

Evaluate (5 minutes)

> **Check-in**
>
> The temperature is 37 °C in Spain in July. How does this compare with body temperature, which is 98.6 °F?

Answer: Converting 37 °C to Fahrenheit: °F = $\frac{9}{5}$(37 °C) + 32 = 98.6 °F. This is identical to body temperature and well above room temperature, so it will feel rather hot.

Homework

Assign the reading and exercises for Weather Lesson 4 in the student text.

Optional: Assign the project Different Thermometers.

LESSON 5 OVERVIEW

Absolute Zero
Kelvin Scale

Lesson Type
Computer Activity: Individuals

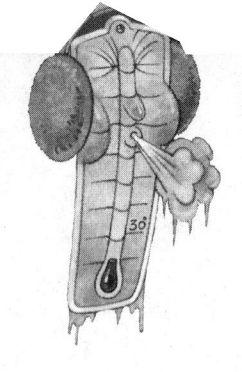

Key Ideas

Celsius and Fahrenheit are the two most commonly used temperature scales. The Kelvin scale is a third temperature scale widely used by scientists. The Kelvin scale is based on the idea that the temperature of a gas should be assigned a value of zero when the volume of the gas is zero. Zero kelvin, referred to as absolute zero, corresponds to −273.15 °C. The average speed of gas particles decreases as temperature decreases.

As a result of this lesson, students will be able to

- describe the relationship between the Celsius and Kelvin temperature scales
- explain the concept of absolute zero
- describe the motion of gas molecules according to the kinetic theory of gases

Focus on Understanding

- Negative temperatures can be confusing to students—especially the idea that, for negative temperatures, as the temperature decreases, the absolute value of the temperature increases.
- Students might find it baffling that Kelvin units and Celsius units are equivalent in size, given their dramatic difference in values.

Key Terms

absolute zero
Kelvin scale
kinetic theory of gases
temperature

What Takes Place

Students complete a worksheet that introduces the Kelvin temperature scale. First, they explore the Celsius scale and contemplate what happens when the volume of a gas drops to zero. The new scale is introduced and compared to other temperature scales. Students observe a simulation depicting the motions of gas particles. Then they explain the expansion and contraction of gases with changing temperature as changes in the average speeds of the particles.

Topic: Chemistry Simulations
Visit: www.SciLinks.org
Web code: KEY-305

Materials

- student worksheet
- transparency—Absolute Zero
- computer and projector to display simulation

Setup

Download a gas properties simulation from SciLinks that allows you to explore the gas laws by varying pressure, temperature, volume, and the number of particles.

Set up the projector to display the simulation and run through the simulation before class.

LESSON 5 GUIDE

Absolute Zero
Kelvin Scale

Engage (5 minutes)

Key Question: How cold can substances become?

> **ChemCatalyst**
>
> Researchers have recorded the temperature on Triton, a moon of Neptune, as −235 °C.
>
> 1. Do you think carbon dioxide, CO_2, would be a solid, a liquid, or a gas at this temperature? Explain your reasoning.
> 2. What do you think is the coldest temperature something can get to? What limits how cold something can get?

Sample Answers: 1. Carbon dioxide would be a solid at this temperature. Everything becomes a solid if it gets cold enough. 2. Some students may say zero degrees, while others may note that there are negative Celsius and Fahrenheit temperatures. They might say there is no limit to how cold something can get. Some might say that the lowest temperature occurs when all motion in a sample of matter stops.

Discuss the ChemCatalyst

➡ Assist students in sharing their initial ideas on cold temperatures.

Sample Questions

- What do you think will happen to gaseous carbon dioxide molecules when the temperature gets as low as −235 °C? Explain your thinking.
- Do you think the molecules move closer together or farther apart as they are cooled?
- What is the lowest temperature you think a substance can reach? Explain your thinking.
- What is the smallest volume you think a gas could occupy? Explain your thinking.

Explore (15 minutes)

Introduce the Activity

➡ Tell students that they will be introduced to a new temperature scale and shown a computer simulation of the motions of gas particles. Students will work together as a class with the teacher on the worksheet.

Guide the Activity

- For Part 2: Computer Activity, follow these steps:
- Select constant volume.
- Add about 20 gas molecules.
- Ask students to observe the motions of the molecules.
- Ask students to observe the motions of the molecules as you raise and lower the temperature.

LESSON 5 ACTIVITY

Absolute Zero
Kelvin Scale

Name _____
Date _____ Period _____

Purpose
To introduce the Kelvin temperature scale and a model describing the motion of gas particles.

Part 1: The Kelvin Scale

1. The volume of a sample of gas was measured at several temperatures. The data are given in the table. Plot the data points on the graph.

Temperature	Volume
10.0 °C	50 mL
50.0 °C	57 mL
100.0 °C	66 mL

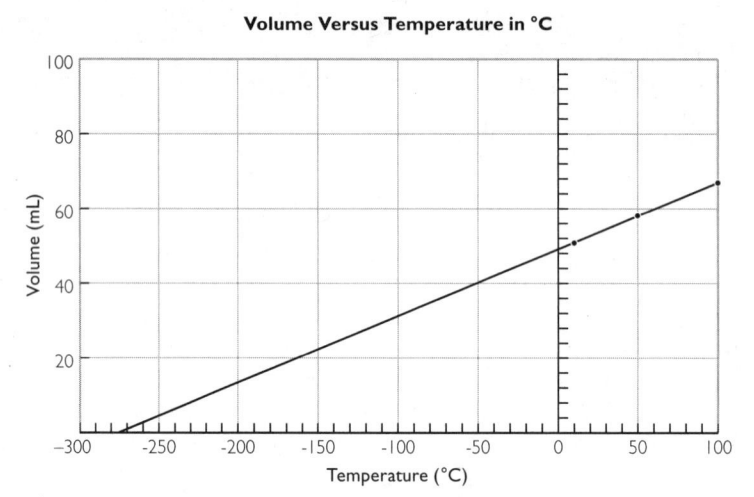

Volume Versus Temperature in °C

2. Draw the best straight line you can through the points on the graph.
3. Use the graph to find the temperature if the volume of this gas decreases to zero.

   approximately −270 °C

4. Do you think the temperature can keep dropping indefinitely? Explain your reasoning.

   It's impossible for the volume of matter to go to zero or below. Thus, the temperature cannot keep dropping.

5. Compare the Fahrenheit, Celsius, and Kelvin thermometers on the next page. Fill in the temperatures in Kelvin that correspond to the temperatures on the Fahrenheit and Celsius thermometers.

6. Zero Kelvin (0 K) is also called **absolute zero.** What is absolute zero equal to in degrees Celsius? in degrees Fahrenheit?

   −273 °C; −459 °F

7. Mark where you would put 0 °F and 0 K on the thermometers.

	Fahrenheit	Celsius	Kelvin
Boiling point H$_2$O	212 °F	100 °C	373 K
Body temp	98.6 °F	37 °C	310 K
Room temp	68 °F	20 °C	293 K
	50 °F	10 °C	283 K
Freezing point H$_2$O	32 °F	0 °C	273 K
	0 °F		
Freezing point mercury (Hg)	−102 °F	−39 °C	234 K
			0 K

Fahrenheit scale Celsius scale Kelvin scale

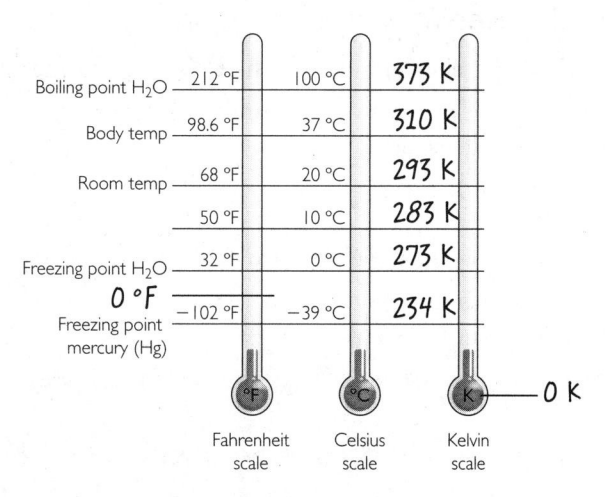

Part 2: Computer Activity

1. Observe the gas particles computer simulation. List at least four features of the model.
   Example: The particles are in constant motion.

   Sample answers: The motion is random. The gas particles move in straight lines. The speed of the gas particles varies. There is a lot of space between gas particles. The particles bounce off the walls or each other.

2. What causes the gas particles to change direction in the model?

   A gas particle changes direction when it collides with another particle or with the wall of the container.

3. What do you notice about the speeds of the particles in the model?

   The particles are not all moving at the same speed.

4. What do you observe when the temperature changes in the model?

   The particles move faster as the temperature increases and slower as the temperature decreases.

5. **Making Sense**   How can you use the motions of the gas particles to explain why gases expand on heating and contract on cooling?

   When the temperature increases, the particles move faster causing them to spread apart, so the volume increases. When the temperature decreases, the molecules move slower causing the volume to decrease.

6. **If You Finish Early**   Which is denser, air at 10 °C or air at 4 °C? Explain your reasoning.

   As the air cools from 10 °C to 4 °C, the mass stays the same but the volume decreases. The gas molecules are closer together, so the value of m/V is larger. Thus the cold air is denser than the warm air.

Explain and Elaborate (15 minutes)

Discuss the Kelvin Scale

- Display the transparency comparing the Celsius scale and the Kelvin scale.
- During the discussion, draw vertical lines showing where 0 °C and 0 K each intersect the opposite scale.

Sample Questions

- Can the volume of a gas be negative? Why or why not?
- What is the lowest temperature that you think can be reached on the Celsius scale? Explain your thinking.
- Do you think it is possible for a substance to reach absolute zero? Why or why not?
- What might be the advantages of the Kelvin scale?

Key Points

On the Celsius scale, the temperature at which the volume of a gas is theoretically equal to 0 is −273 °C. (The actual value is −273.15 °C.) If you shift the Celsius temperature scale to the left by 273° by adding 273 to each temperature, the result is a scale on which the temperature will be zero when the volume is zero. This new temperature scale is called the Kelvin scale. The unit of temperature on the Kelvin scale is the kelvin (K). The symbol ° and the word *degree* are not used with the Kelvin scale. For example, it is correct to say 5 kelvins, not 5 degrees Kelvin.

$$K = °C + 273$$

$$°C = K - 273$$

A temperature of 0 K is referred to as absolute zero. This is considered to be the lowest temperature that could hypothetically be reached. Absolute zero has never been attained in the laboratory. As of 2003, the lowest temperature reached in a laboratory was 0.0000000005 K (500 picokelvins). The lowest temperature ever recorded on earth is −89 °C, in Antarctica. This is equivalent to 184 K. The lowest

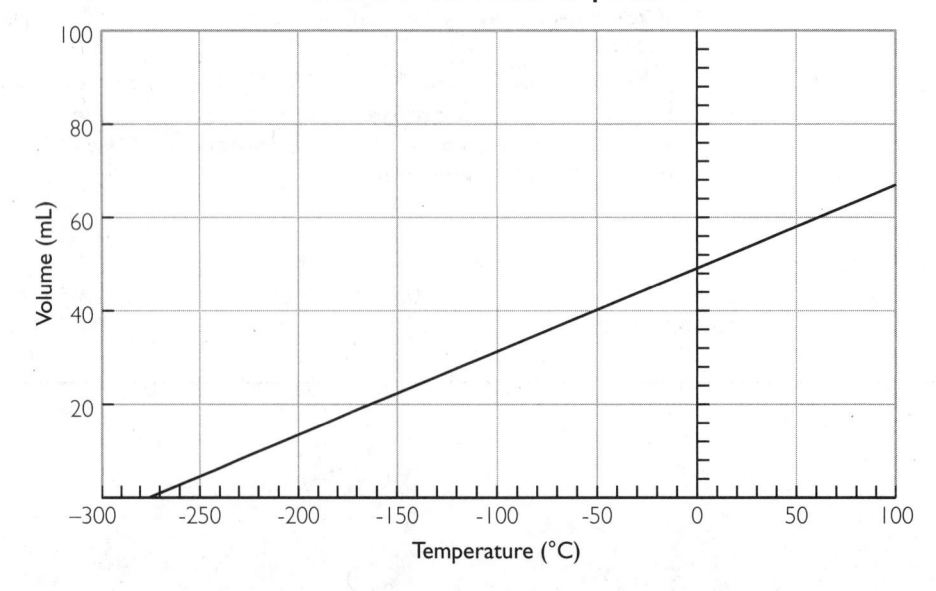

**Celsius Scale
Volume of Gas Versus Temperature**

temperature recorded in the solar system was on Triton, a moon of Neptune, and was reported to be −235 °C, or 38 K. At this temperature, the surface of Triton is thought to consist of oceans of nitrogen and glaciers of methane.

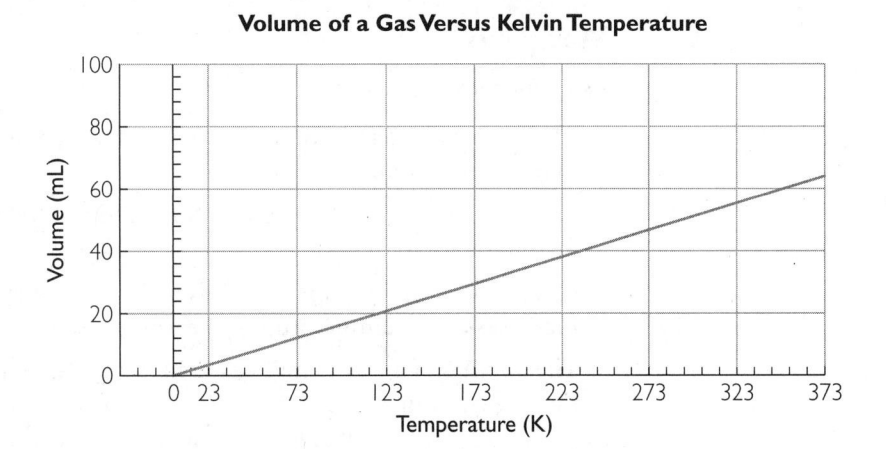

Volume of a Gas Versus Kelvin Temperature

Discuss the Computer Activity

⇒ Run the computer simulation again during the discussion portion of the class.
⇒ To help students see the distribution of speeds, run the simulation at a low temperature.

Sample Questions

- Describe the motions of the gas particles.
- What happens to the gas particles as they are heated?
- If the container walls were flexible, what would happen?
- How can you explain the change in volume of a gas that is heated or cooled, in terms of the motions of particles?
- What do you think would happen to the motions of the gas particles if you could cool a gas to 0 K?

Key Points

The model displayed in the simulation is the kinetic theory of gases. This theory maintains that gas particles exhibit these features:

- Gas particles are constantly moving.
- The motion of gas particles is random.
- Gas particles move in straight lines.
- The speeds of the particles are not all the same.
- Gas particles have a lot of space to move around in. (They are tiny compared to the space they are found in.)
- Gas particles change directions when they hit each other or the walls of the container.

As the temperature increases, the average speed of the particles increases. At higher temperatures, the gas particles move faster, and you can imagine that they bounce harder off the walls. At lower temperatures, the particles move slower, and you can imagine that they hit the walls with less force. Indeed, if the container were flexible, like a balloon, the hot gas particles would expand the balloon. The balloon would shrink if the gas particles were cooled.

The temperature of a gas is a measure of the average energy of motion of the gas particles. Scientists hypothesize that if it were possible to reach absolute zero, the motions of atoms and particles would stop.

Wrap-up

Key Question: How cold can substances become?

- The Kelvin scale assigns a value of zero to the temperature of a gas with a hypothetical volume of zero.
- Conversion between the Celsius scale and the Kelvin scale is given by the formula K = °C + 273.
- Gas particles are in constant motion. The average speed of gas particles increases as the temperature increases.
- The volume of a gas changes as the temperature changes, if the gas is in a flexible container.

Evaluate (5 minutes)

Check-in

1. Describe three features of the motions of gas particles.
2. Use the motions of gas particles to explain why gases expand when they are heated.

Answers: 1. Sample answers: Gas particles are in constant, random straight-line motion. Gas particles move at different speeds and generally do not interact with one another because they have a lot of space to move around in. Gas particles change directions when they hit the walls of the container or each other. 2. When gases are heated, the gas particles move faster. They push harder on the walls of the container. If the container has a flexible size, its size will increase.

Homework

Assign the reading and exercises for Weather Lesson 5 in the student text.

LESSON 6 OVERVIEW

Sorry, Charlie
Charles's Law

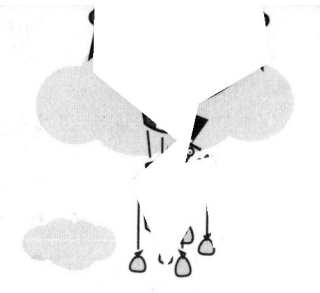

Lesson Type
Classwork:
  Pairs
Demo (optional):
  Pairs

Key Ideas

The mathematical relationship between volume and temperature for a certain amount of gas at constant pressure is known formally as Charles's law. It is expressed mathematically as $V = kT$ or $k = V/T$, where k is the proportionality constant. This proportionality holds true only if the Kelvin scale is used. If one set of values for volume (V) and temperature (T) is known for a specific gas sample, it is possible to solve for any value of V or T for that gas sample. The value of the proportionality constant, k, depends on the amount of gas and the pressure of the gas. If the pressure or the amount of gas changes, the proportionality constant changes.

As a result of this lesson, students will be able to

- explain Charles's law and use it to solve simple gas law problems involving volume and temperature
- explain two methods for determining the volume of a gas if its temperature is known

Focus on Understanding

- The conditions under which the proportionality constant, k, changes sometimes confuses students. The value of k stays the same for a given sample of gas. The value of k changes if the amount of gas or its pressure is changed.
- Students have not yet been introduced to the concept of pressure, so we avoid direct mention of it in the classwork. When defining Charles's Law in the Explain and Elaborate discussion, we say that the pressure does not change rather than say that the pressure is constant.
- Throughout the unit, we avoid using the term *constant* to mean unchanging or fixed to avoid confusion with the proportionality constant.

SciLinks
Topic: Gas Laws
Visit: www.SciLinks.org
Web code: KEY-306

Key Term

Charles's law

What Takes Place

In this lesson, students investigate the mathematical relationship between the volume and temperature of a gas. Teachers may complete an optional demo using liquid nitrogen that allows students to observe a change in the volume of a gas in response to a change in temperature. Students then complete a worksheet that

Section I

supports them in reasoning and in performing calculations using Charles's law. Finally, students solve gas law problems by finding volume and temperature using the value of the proportionality constant, $k = V/T$.

Materials

- student worksheet
- transparency—Balloon Volume Versus Temperature
- Math Card: Charles's law, for each student (optional)
- lava lamp for ChemCatalyst (optional)

Demonstration materials (optional)

- 2 small balloons
- tape measure
- liquid nitrogen
- hot water

LESSON 6 GUIDE

Sorry, Charlie
Charles's Law

Engage (5 minutes)

Key Question: How can you predict the volume of a gas sample?

> **ChemCatalyst**
>
> A lava lamp contains a waxy substance and water, which do not mix, and a light bulb at the base. As the bulb heats the waxy substance, it rises. Near the top of the lamp, the waxy substance cools and falls. Explain why this happens.

Sample Answer: Students might not have a complete answer at this point. They might say that the wax expands as it is heated and contracts as it cools. The hotter wax occupies a larger volume and therefore is less dense. Substances that are less dense float on top of denser substances.

Discuss the ChemCatalyst

Sample Questions

- How does the density of a substance change as it is heated?
- Why does the wax in the lava lamp rise and fall?
- Explain why heating vents in a room are near the floor and not on the ceiling.
- What do you think happens to air warmed by the Sun during the day?
- How do you think the expansion and contraction of gases affect the weather?

Explore (15 minutes)

Complete the Demonstration (optional)

➡ Demonstrate the change in volume of a balloon as it is heated or cooled.

1. Blow up a small balloon with an elongated shape.
2. Measure the length of the balloon. Write this length on the board.
3. Cool the balloon in a container filled with liquid nitrogen. Measure the length of the balloon when it is cold.
4. Warm the balloon in hot water. (Be careful that the balloon does not melt!) Measure the length of the balloon.
5. Ask students to speculate how they could determine the exact volume inside each balloon.

Introduce the Classwork

➡ Ask students to work in pairs on the worksheet. Tell them they will be exploring how the volume of a gas is related to its temperature.

LESSON 6 — CLASSROOM
Sorry, Charlie
Charles's Law

Name _____
Date _____ Period _____

Purpose
To calculate changes in the volume of a gas as they relate to temperature changes.

Questions

1. A Happy Birthday balloon is filled with three breaths of air. It has an initial volume, V_1, of 400 mL at the initial temperature, T_1, of 285 K. The air in the balloon is cooled to 265 K and the volume decreases. Next, the air in the balloon is heated to 300 K and the volume increases. Calculate the missing values for the birthday balloon.

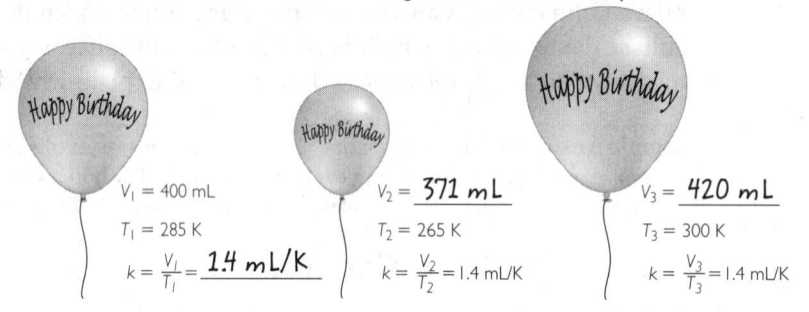

   Balloon 1: V_1 = 400 mL; T_1 = 285 K; $k = \frac{V_1}{T_1}$ = __1.4 mL/K__

   Balloon 2: V_2 = __371 mL__; T_2 = 265 K; $k = \frac{V_2}{T_2}$ = 1.4 mL/K

   Balloon 3: V_3 = __420 mL__; T_3 = 300 K; $k = \frac{V_3}{T_3}$ = 1.4 mL/K

2. A New Year's balloon is inflated with five breaths of air. It has an initial volume of 600 mL at 285 K. It is heated to a temperature that changes the volume of air in the balloon to 630 mL. Next the air in the balloon is cooled to 265 K. Calculate the missing values for the New Year's balloon.

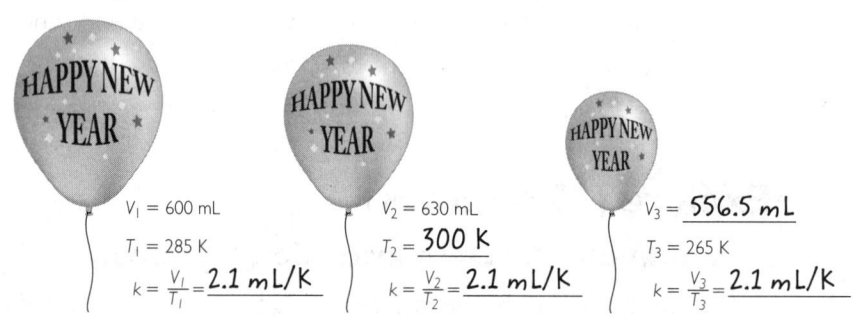

   Balloon 1: V_1 = 600 mL; T_1 = 285 K; $k = \frac{V_1}{T_1}$ = __2.1 mL/K__

   Balloon 2: V_2 = 630 mL; T_2 = __300 K__; $k = \frac{V_2}{T_2}$ = __2.1 mL/K__

   Balloon 3: V_3 = __556.5 mL__; T_3 = 265 K; $k = \frac{V_3}{T_3}$ = __2.1 mL/K__

3. The volume of a sample of gas is proportional to its temperature in kelvins but the volume is *not* proportional to its temperature in degrees Celsius. Use data from Question 1 to provide evidence to support this assertion.

   T_1 = __12__ °C T_2 = __−8__ °C T_3 = __27__ °C

   $k = V_1/T_1$ = __33__ mL/°C $k = V_2/T_2$ = __46__ mL/°C $k = V_3/T_3$ = __16__ mL/°C

   For each temperature in degrees Celsius, you need to multiply by a different value of k in order to obtain the volume.

4. Plot volume versus temperature in kelvins for the Happy Birthday balloon and the New Year's balloon on the graph. Label each line.

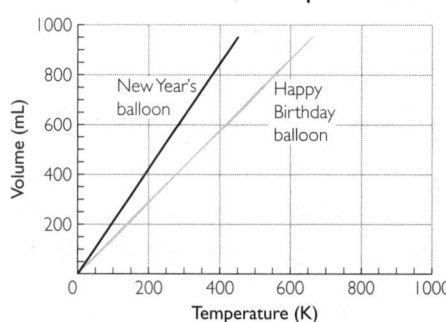

Volume Versus Temperature

[Graph showing Volume (mL) versus Temperature (K), with New Year's balloon line (steeper) and Happy Birthday balloon line]

5. Use the graph to find the approximate volume of the Happy Birthday balloon and the New Year's balloon at a temperature of 400 K.

Happy Birthday balloon: ~560 mL
New Year's balloon: ~840 mL

6. Why do you suppose the Happy Birthday balloon has a different proportionality constant, k, than the New Year's balloon?

The change in volume with temperature is different because the balloons have different amounts of gas in them.

Problem Solving

(Remember to *always* convert temperatures to the Kelvin scale.)

1. The beginning volume of a gas is 500 mL at 20 °C. The temperature is raised to 35 °C. What is the new volume of the gas?

Convert to Kelvin: 20 °C = 293 K, 35 °C = 308 K.
Find k: k = V/T = 500 mL/293 = 1.7 mL/K
V = (308 K)(1.7 mL/K) = 524 mL.

2. **Making Sense** Suppose you have a Valentine's Day balloon with a volume of 300 mL at 300 K.

   a. Is the proportionality constant larger or smaller than that for the birthday balloon?

   k = V/T = 1.0 mL/K, which is smaller than the value
   k = 1.4 mL/K for the birthday balloon.

   b. At the same temperature, which balloon is smaller, the Valentine's Day balloon or the New Year's balloon?

   The Valentine's Day balloon, because V = k · T and the value of k is smaller.

3. **If You Finish Early** On the graph of volume versus temperature for a gas, what does the slope of the line relate to?

   The slope of the line is equal to the proportionality constant. It relates to the ratio of V to T.

Explain and Elaborate (15 minutes)

Introduce Charles's Law

[T] ⟶ Display the transparency Balloon Volume Versus Temperature.

Sample Questions

- How can you use the value of k to determine the volume of a gas sample at different temperatures?
- How does the value of k change with the size of the gas sample?
- How does the value of k relate to the steepness of the line on the graph?
- Why is it necessary to measure only one volume and temperature in order to draw a graph of volume versus temperature for a gas?

Key Points

Charles's law states that volume, V, is proportional to the Kelvin temperature, T. This means that the volume is equal to the proportionality constant, k, times the temperature: $V = kT$. This relationship was described first by the French scientist Jacques Charles in 1802. It holds true only if the amount of gas and the gas pressure don't change.

> **Charles's law:** For a given sample of gas at a certain pressure, the volume of gas is directly proportional to its Kelvin temperature.

The proportionality constant, k, indicates how much the volume of a gas changes per kelvin. You can calculate the value of k from one measurement of volume and temperature for a gas sample: $k = V/T$. Then you can multiply a different temperature by k to find the volume at this temperature. This is true only if temperatures are given in the Kelvin scale.

Because volume is proportional to temperature, the graph of volume versus temperature for a gas sample is a straight line that goes through the origin, (0, 0). Each point on the line is a volume-temperature pair that satisfies the equation $V = kT$. You can use the graph to determine the volume at any temperature. The value of k is equal to the slope of the line and is different for different quantities of gas. This means that the lines will vary in steepness for balloons with different quantities of gas inside.

Volume Versus Temperature

Living By Chemistry Teacher Guide — Unit 3 Weather

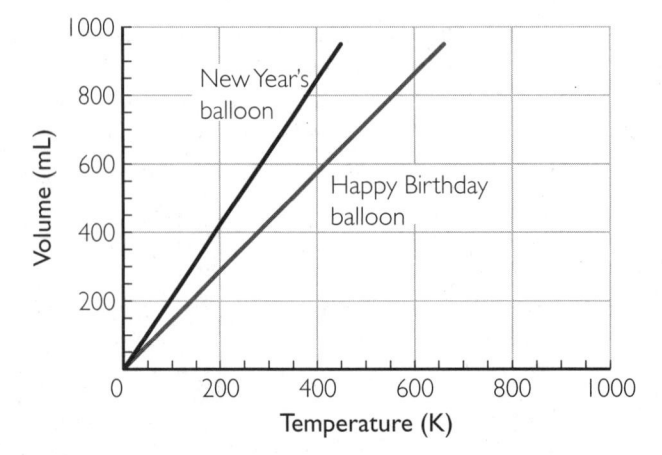

Complete an Example Using Charles's Law

→ Hand out Math Card: Charles's Law (optional) and help students relate V, T, and k.

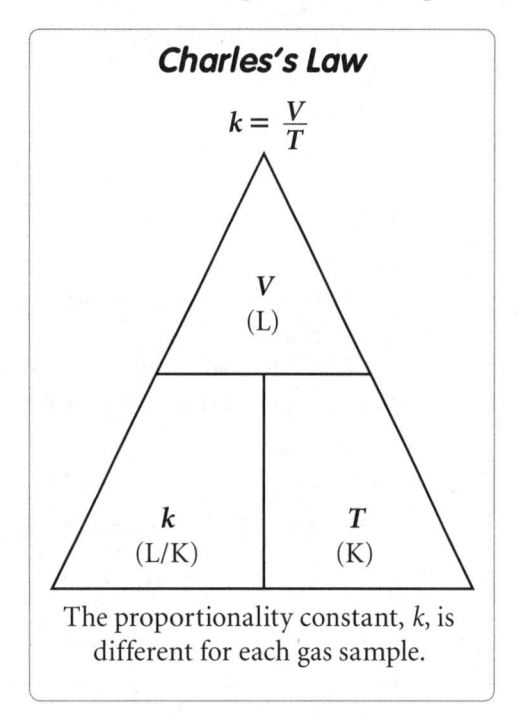

Charles's Law

$$k = \frac{V}{T}$$

V (L)

k (L/K) T (K)

The proportionality constant, k, is different for each gas sample.

Example

The first thing in the morning, you fill a balloon with air to a volume of 180 mL at 50 °C. After several hours out in the Sun, the air inside the balloon has warmed to 85 °C. Calculate the new volume of the balloon.

Solution

Step 1: Predict whether the volume will increase or decrease.

The temperature increases, so the volume of the balloon must increase.

Step 2: Convert all temperatures to the Kelvin scale:

$$T_1 = 50 °C + 273$$
$$= 323 \text{ K}$$

$$T_2 = 85 °C + 273$$
$$= 358 \text{ K}$$

Step 3: Find the proportionality constant, k.

Use the volume and temperature when you first inflated the balloon to determine the value of k:

$$k = \frac{V}{T} = \frac{180 \text{ mL}}{323 \text{ K}} = 0.56 \text{ mL/K}$$

Step 4: Apply Charles's law. Use the value of k to find V at the new temperature:

$$V = kT = (0.56 \text{ mL/K})(358 \text{ K}) = 200 \text{ mL}$$

Step 5: Check that the answer makes sense and matches prediction.

The volume has increased from 180 mL at 50 °C to 200 mL at 85 °C. Thus, the volume has increased slightly, as expected for a small increase in temperature.

Discuss Why Hot Air Rises

Sample Questions

- What happens to the density of a gas as the temperature increases?
- Suppose a balloon is heated such that the density is lower than that of the surrounding air. Do you expect the balloon to float or sink? Explain.
- Propose two ways to decrease the density of the gas inside a balloon.

Key Point

Hot air rises because it is less dense than cooler air. When the temperature of a mass of air increases, the molecules move faster and the volume increases. Because density is given by m/V, the density of the air decreases when V gets larger. Less dense substances float on denser substances, so a larger warm air mass floats above a larger cooler air mass.

Wrap-up

Key Question: How can you predict the volume of a gas sample?

- You can use the proportionality constant to calculate the volume of a gas sample at any temperature if the temperature is in kelvins and the amount of gas and the gas pressure don't change: $V = kT$. This is known as Charles's law.
- The proportionality constant, k, depends on the amount of gas in a sample and the pressure of the gas.
- As the temperature increases, the volume of a gas sample increases and the density of the gas decreases. This is why warmer gas rises above cooler gas.

Evaluate (5 minutes)

> **Check-in**
>
> A sample of gas has a volume of 120 L at a temperature of 40 °C. The temperature drops to −10 °C. If nothing else changes, what is the new volume of the gas?

Answer: The temperature decreases, so the volume decreases. Convert the temperatures to the Kelvin scale. Then determine the proportionality constant. $k = V/T = 120$ L$/313$ K $= 0.38$ L/K. Now use the value of k to determine the new volume at -10 °C (263 K). $V = kT = (0.38$ L/K$)(263$ K$) = 100$ L.

Homework

Assign the reading and exercises for Weather Lesson 6 in the student text.

LESSON 7 OVERVIEW

Front and Center
Density, Temperature, and Fronts

Lesson Type
Activity:
 Whole class or
 Groups of 4
Demo:
 Whole class

Key Ideas

Weather fronts, which occur where two air masses meet, are responsible for the majority of storms that occur. A cold front forms when a cold air mass catches up with a warm air mass, while a warm front forms when a warm air mass catches up with a cold air mass. When warm and cold air meet, density differences force the warm air upward, and clouds form.

As a result of this lesson, students will be able to

- explain the roles of temperature and density in the movement of cold and warm air masses
- describe the weather patterns associated with warm fronts and cold fronts

Focus on Understanding

- Air mass may seem odd terminology to students. It simply refers to a large mass of air that is consistent with respect to temperature and moisture.

Key Term

air mass

What Takes Place

This lesson summarizes what has been learned about the weather so far and integrates the chemistry learning with the context. Students observe a demonstration at the beginning of class that shows water masses of two different temperatures and densities mixing, to simulate the meeting of air masses. Weather variables are re-examined using the color transparencies from Lesson 1: Weather or Not. Students complete a worksheet to explore the interaction between weather fronts and precipitation.

Materials

- student worksheet
- transparencies—ChemCatalyst, Fronts, base map

Per group of 4 (or one class set)

- 7 weather map color transparencies (from Lesson 1)—(Jet Stream, Temperature Highs, Cloud Cover, Fronts, Precipitation, Air Pressure, and blank base map)

Demonstration materials

- clear plastic tank with partition
- red and blue food coloring
- warm water (~30 °C) and room temperature water (~20 °C)

Section I   Lesson 7 Front and Center   55

Setup

We recommend that you complete at least one practice run with the demonstration apparatus before class. The apparatus is a clear, flat acrylic tank divided into two equal volumes with a removable partition. The key to success is to remove the partition at a speed that facilitates the slow movement of one body of colored water over the other.

While students are working on the ChemCatalyst, add cold water (20 °C) and blue food coloring to one chamber in the tank. Add hot water (30 °C) and red food coloring to the other chamber.

If you have several sets of color transparencies, students can do Part 1 on the worksheet in groups. If you have one set, you can use an overhead projector and do Part 1 as a whole-class discussion. Superimpose the transparencies as needed for each question.

Cleanup

Water with food coloring can be disposed of in the sink. Rinse out the tank in preparation for the next class.

LESSON 7 GUIDE

Front and Center
Density, Temperature, and Fronts

Engage (5 minutes)

Key Question: How do weather fronts affect the weather?

ChemCatalyst

Large air masses form over different regions of land and ocean. These air masses have a consistent temperature and moisture content.

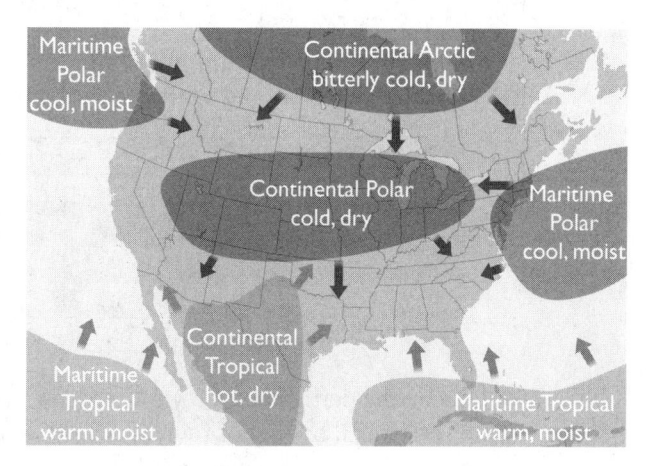

1. What patterns do you notice in the temperatures and moisture content of the air masses shown on the map?
2. Why do you think clouds form when the Continental Polar air mass collides with the Maritime Tropical air mass?
3. Use the concept of density to explain why warm air in the Maritime Tropical air mass rises, while cold air in the Continental Polar air mass descends.

Sample Answers: 1. Cold air masses are in the north. Warm air masses are near the equator. Air masses over water have higher moisture content than those over land. 2. A tropical air mass has a lot of moisture. When it meets a polar air mass, the temperature decreases. The water condenses to form clouds. 3. The denser, cool air descends. The less dense, warm air rises.

Discuss the ChemCatalyst

Sample Questions

- Describe an air mass.
- Why is warm air less dense than cold air?
- What do you predict will happen when a warm air mass collides with a cold air mass?
- Explain why warm air rises and cold air descends.

Section I   Lesson 7 Front and Center   57

Explore (20 minutes)

Complete the Demonstration

⟹ Add cold water (20 °C) and blue food coloring to one chamber in the tank. Add hot water (30 °C) and red food coloring to the other.

⟹ Use the plastic chamber and colored water to simulate what happens when two air masses of differing temperatures meet each other. Done properly, the mixing water masses resemble a weather front that forms at the boundaries of two air masses.

- Tell students that the red water represents a tropical air mass and the blue water represents a polar air mass.
- Ask students to predict what will happen if you remove the partition dividing the two chambers.
- Hold a white paper behind the container so that the interaction between the two solutions can be seen more clearly.
- Carefully remove the center dividing wall, allowing the two solutions to interact.
- Ask students to describe what they observe and to offer explanations for their observations.
- Bring the class's attention back to the apparatus when the warm and cold water samples have layered vertically.

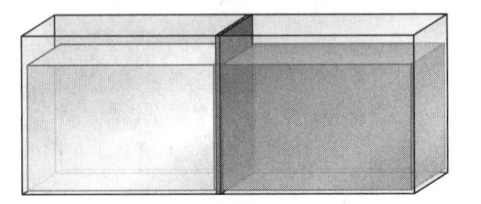

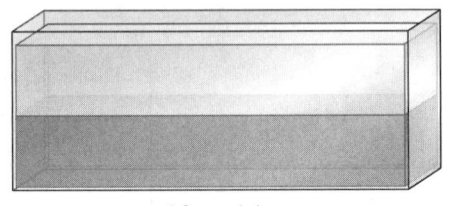

Before mixing   After mixing

⟹ Review the definition of a weather front. Draw the symbols for the weather fronts on the board, by way of review. A cold front forms when a cold air mass catches up with a warm air mass, while a warm front forms when a warm air mass catches up with a cold air mass.

▲▲▲ Cold front   ⌒⌒⌒ Warm front

Introduce the Activity

⟹ Pass out worksheets. Place the first set of transparencies on the overhead, or hand out a set to each group.

LESSON 7 ACTIVITY

Front and Center
Density, Temperature, and Fronts

Name _____
Date _____ Period _____

Purpose
To investigate how fronts affect the weather.

Part 1: Weather Maps
Reexamine the weather maps from Lesson 1 to answer the questions.

1. Examine the Fronts Map, Cloud Cover Map, and Precipitation Map together. What relationships do you see among fronts, clouds, and precipitation?

   Most of the precipitation on the map occurs in the areas around warm and cold fronts. There are clouds over the fronts.

2. Where would you expect to see warm and cold air masses on the Fronts Map? Draw them on this map.

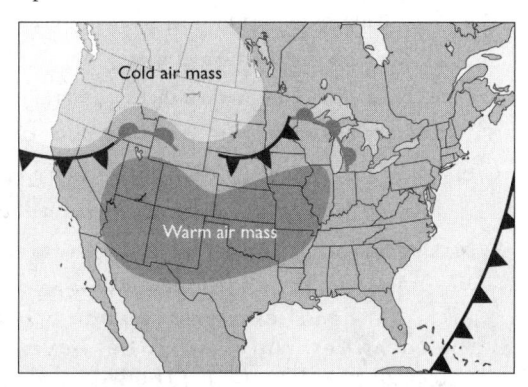

Part 2: Warm and Cold Fronts

1. Why is a cold air mass denser than a warm air mass?

   A denser air mass has more gas molecules per unit of volume. It has more matter in the same space. Cold air contracts, so the molecules are closer together.

2. Explain why clouds might form when a warm air mass collides with a cold air mass.

   The moisture in the warm air mass will condense as it cools, forming clouds.

3. Examine the illustration showing what happens at a cold front.

   Cold front: Cold air overtakes warm air.

   a. Explain why warm air is pushed up by the cold front.

    Warm air is less dense than cold air. It will float on cold air, so the cold air moves in underneath while the less dense warm air rises.

   b. Where do clouds form when there is a cold front?

    Clouds form behind and in the area of an advancing cold front.

   c. Where does precipitation fall when there is a cold front?

    Precipitation falls behind and in the area of a cold front.

4. Examine the illustration showing what happens at a warm front.

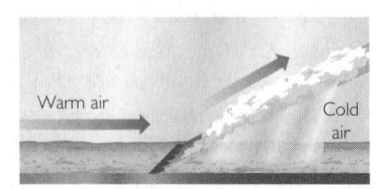

Warm front: Warm air overtakes cold air.

   a. What happens to the warm air when it overtakes the cold air?

    It rises above the cold air.

   b. Where do clouds form when there is a warm front?

    Clouds form in front of the advancing front.

   c. Where does precipitation fall when there is a warm front?

    Precipitation falls in front of an advancing warm front.

5. Making Sense What does air density have to do with weather fronts?

Weather fronts are the areas where air masses of significantly different densities overtake each other. Cold air masses have high densities. Warm air masses have low densities. In areas where warm air meets cold air, the warm air is less dense, so it rises above the cold air.

6. If You Finish Early Nearly eighty percent of the air in our atmosphere is nitrogen gas, N_2, while water vapor makes up only 1% of the air. Why doesn't it rain liquid nitrogen instead of rainwater?

Nitrogen condenses into a liquid at a much cooler temperature than that found on Earth. (In fact, nitrogen does not condense from a gas into a liquid until it reaches 77 K, or −196 °C or −321 °F.)

Explain and Elaborate (15 minutes)

Discuss Weather Fronts

[T] ⇒ Use the transparencies Fronts and base map to assist the class in processing what they have learned.

Sample Questions

- What differences do you see between cold fronts and warm fronts?
- What is the relationship between weather fronts and precipitation?
- How are fronts and air masses related?
- How does density influence weather fronts? How does temperature influence weather fronts?

Key Points

Fronts occur between the boundaries of warm and cold air masses.   In general, warm (tropical) air masses move up across the North American continent from the south, and cold (polar) air masses move down across the continent from the north. The shape of a front tells you something about where the cold or warm air masses lie. You can see that the cold fronts are bowed down and the warm fronts are bowed up on the map of North America.

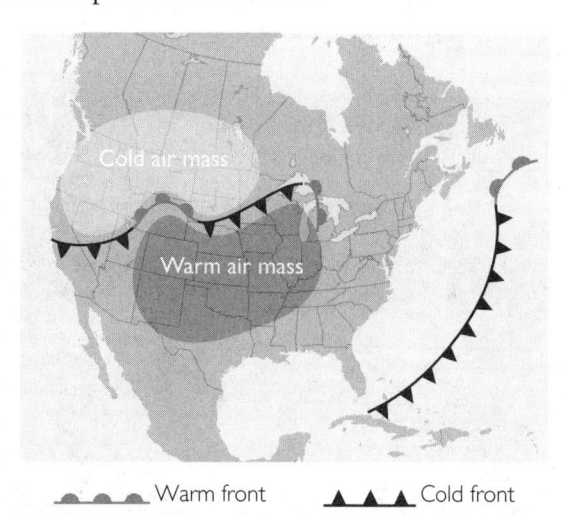

⌢⌢⌢ Warm front   ▲▲▲ Cold front

Warm and cold air masses have different densities.   Warm air masses rise above cold air masses because they are less dense. When this happens, clouds form. The cold temperatures cause water in the warm air mass to change phase from a vapor to a liquid. This is why fronts are associated with storms and rain.

Cold front. Cold air overtakes warm air.

Warm front. Warm air overtakes cold air.

The weather associated with cold and warm fronts differs. The clouds that form with cold fronts tend to be thicker, puffy clouds similar to those seen with thunderstorms. The clouds form quickly, directly in the area of the cold front. The clouds associated with warm fronts are thinner and not as puffy. In advance of warm fronts, you may have days of clouds before the rain actually arrives. Precipitation tends to *occur at or just behind* cold fronts and come *in advance of* warm fronts.

Summarize the Weather Unit So Far

➟ Use the base map transparency to review the different weather concepts learned so far.

➟ Draw one warm front and one cold front on the map, as shown. Ask students where to draw clouds, precipitation, highs, lows, and the jet stream in the appropriate places.

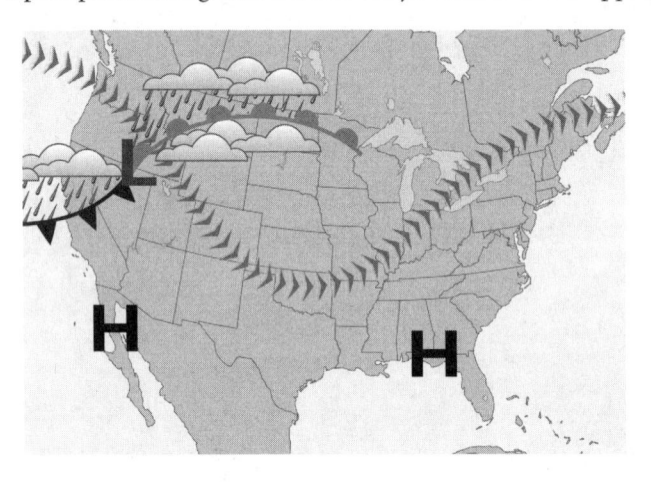

➟ Ask students to provide their reasoning for what is placed on the map.

Sample Questions

- What are some generalizations you can make about weather and the different variables that affect weather?
- If you were looking at a map that showed only weather fronts, where would you predict precipitation?
- What would you know about air masses from such a map?
- What would you know about cloud cover?
- What would you know about the location of lows and highs?

Key Points

Interactions among the temperature, volume, and density of air masses contribute significantly to the formation of weather. Here are some generalizations about weather:

- Fronts occur at the boundaries between warm and cold air masses. Warm air overtaking cold air is called a warm front. Cold air overtaking warm air is called a cold front.
- Warm air, which is less dense, layers over the denser cold air.
- Clouds and steady light rain form ahead of a warm front. Clouds and heavy showers form at and behind a cold front.
- On weather maps, Ls are closely associated with fronts while Hs appear away from the fronts. Highs are associated with clear skies. Lows are associated with storms and cloudy skies.

Wrap-up

Key Question: How do weather fronts affect the weather?

- Fronts form where air masses overtake each other.
- Storms and precipitation occur at both warm fronts and cold fronts.
- Density differences account for the movement and layering of air masses when they meet.
- Temperature differences account for phase changes that occur when air masses meet, causing precipitation.

Evaluate (5 minutes)

Check-in

A warm front is approaching your hometown. It is one day away. What would you expect to observe in the way of weather?

Answer: Clouds and rain form in front of an advancing warm front. There will be increasing clouds and steady and light rainfall.

Homework

Assign the reading and exercises for Weather Lesson 7 and the Section I Summary in the student text.

II Pressing Matter

Section II of the Weather unit consists of seven lessons focusing on gas pressure. Lesson 8 is a lab that provides students firsthand experience with gas pressure and density. In Lesson 9, students observe a series of demonstrations featuring the effects of gas pressure on the world around them. Lesson 10 introduces Boyle's law. In Lesson 11, students investigate flexible and rigid containers and their effects on gas samples. Computer simulations allow students to explore the kinetics of gas behavior under different conditions in Lesson 12. The combined gas law is introduced in Lesson 13 as students follow the progress of a weather balloon through a changing atmosphere. In Lesson 14, students simulate the creation of a cloud inside a bottle and connect their learning to the weather associated with high- and low-pressure systems.

In this section, students will learn

- about the density of gases
- a particulate view of gas pressure
- the relationships among the pressure, volume, and temperature of gases
- Boyle's law, Gay-Lussac's law, and the combined gas law
- weather associated with high- and low-pressure systems

LESSON 8 OVERVIEW

It's Sublime
Gas Density

Lesson Type
Lab:
   Groups of 4

Key Ideas

When a substance changes phase from a gas to either a liquid or a solid, there is a dramatic change in density. This is because the particles in that substance are much farther apart in the gas phase than are those same particles in the solid or liquid phase. When solid carbon dioxide, $CO_2(s)$, changes phase, it goes directly from a solid to a gas. This type of phase change is called sublimation. Solid CO_2 is called "dry ice" because it sublimes directly to a gas without passing through the liquid phase.

As a result of this lesson, students will be able to

- describe the density differences that occur during phase changes
- explain how moisture gets into the atmosphere
- calculate the density of a gas from mass and volume measurements

Focus on Understanding

- Students often believe that they can see water vapor because they confuse it with fog and what most people refer to as steam. Fog and the mist above boiling water (what most people call steam) consist of visible liquid water droplets. We can't see water vapor because it is a colorless gas.
- Students may have difficulty extrapolating from dry ice, which sublimes, to the evaporation of water; however, the magnitudes of the density changes are similar.
- Students might struggle when asked to provide evidence to support molecular views of gases. Encourage them to use their experiences to provide evidence.

Key Terms

sublimation
evaporation

What Takes Place

In this class, students observe a demonstration of the properties of dry ice. After the demonstration, they perform a lab in which they observe the dramatic change in density associated with a phase change from a solid to a gas. At the beginning of class, teams of students quickly measure the mass of a sample of dry ice and place it in a sealed garbage bag to sublime. This step is completed before the ChemCatalyst so the dry ice will have enough time to sublime completely before the end of class. Students complete worksheet questions before measuring

Section II   Lesson 8 It's Sublime

the volume of their bags of CO_2 gas. They then calculate the density of carbon dioxide gas from their data. Sketches of molecular views assist students in conceptualizing gases.

Materials

- student worksheet
- transparency—Dry Ice Setup
- polystyrene foam cooler or ice chest

Demonstration materials

- hot plate
- ice cube
- small samples of dry ice
- 2 small beakers (for dry ice and ice cube)
- small beaker with water
- small beaker with oil (mineral oil or clear cooking oil)
- oven mitt or tongs

Per group of 4

- 5–20 g dry ice
- balance
- 5 gal plastic garbage bag
- twist tie
- polystyrene foam cup
- cardboard box roughly 1 ft by 1 ft by 1 ft (about 18,000 cm^3, close to 5 gal)
- ruler (preferably with the zero at the end, with no offset)

Setup

Before class, acquire 5–7 lb of dry ice. Suppliers are listed in your local phone directory. Before class, use a hammer to crush the dry ice into a powder.

> ### Safety Instructions
> ⚠ Do not touch dry ice with your bare hands. Use gloves or tongs.

Keep the dry ice in a polystyrene foam cooler or ice chest.

Cleanup

You can ask students to "release" the carbon dioxide gas after the lesson and to fold the garbage bags for reuse by the next class.

LESSON 8 GUIDE

It's Sublime
Gas Density

Introduce the Lab (5 minutes)

- Tell students that they will be setting up the first part of their experiment before the ChemCatalyst. Emphasize to students that they must work quickly during this setup.
- Distribute a range of amounts of crushed dry ice in polystyrene foam cups to the groups—somewhere between 5 and 20 g of dry ice.
- Distribute the larger amounts of dry ice first so they will have time to sublime.
- Show the transparency Dry Ice Setup to quickly explain the procedure. Ask students to work in groups of four. The mass of the solid dry ice changes continuously as it sublimes, so it is important to get the dry ice into the garbage bags quickly.

1. One student from each group should get a polystyrene foam cup filled with 5–10 g of dry ice from the teacher and a 5-gallon plastic garbage bag.
2. Remove all the air from the 5-gallon garbage bag.
3. Find the mass of the cup and the dry ice together.
4. Quickly pour the dry ice into the deflated bag and close it tightly so the bag does not leak. Be careful to keep air out.
5. Weigh the empty cup. Subtract this weight from the mass of the cup containing the dry ice in order to determine the mass of the dry ice.

Engage (5 minutes)

Key Question: How do the densities of a solid and a gas compare?

> **ChemCatalyst**
>
> Water exists in many forms, including water vapor, fog, clouds, and liquid water.
>
> 1. Why do you think you cannot see water vapor in the air?
> 2. How are fog and clouds different from water vapor?
> 3. Why do you think airplanes can fly through clouds?

Sample Answers: 1. Some students will know that water vapor cannot be seen because it is a clear colorless gas, with the water molecules very far apart from one another. 2. Many students will know that clouds and fog consist of tiny droplets of water collected together, while water vapor consists of gaseous water molecules distributed throughout other air molecules. 3. Students might say a cloud is only water droplets and air, or a cloud is not very dense.

Discuss the ChemCatalyst

- Assist students in sharing their ideas about water vapor, clouds and fog.

Sample Questions
- Can you see gases in the air? Explain your thinking.
- Why can solid objects move easily through gases?
- Which phase of water is responsible for the composition of clouds?
- Explain the difference between water vapor, fog, and liquid water.
- How do the densities of water vapor and clouds compare to the density of liquid water?

Explore (15 minutes)

Demonstrate Sublimation

➡ The purpose of this demonstration is to introduce the concept of sublimation and to discuss the identity of the "fog" that appears around the dry ice.

➡ This demonstration requires a hot plate set to medium heat, a small beaker of water, a small beaker of mineral or clear cooking oil, an ice cube, and some dry ice.

1. Show students a beaker containing a piece of dry ice and another beaker containing an ice cube (frozen water). Ask students to describe any differences they observe between the dry ice and the water ice.

2. Ask students to predict what will happen if you put the beaker with the water ice on the hot plate. Complete this step. (The water ice will melt, and there will be a hissing sound as the liquid evaporates. The water ice moves around as it "floats" on the liquid water beneath it. Fog is visible.)

3. Ask students to predict what will happen if you put the beaker with dry ice on the hot plate. Complete this step. (The dry ice will slowly disappear. A liquid is not visible, and the dry ice does not move around. You do see some fog—this is from water vapor in the air that condenses.)

4. Ask students what they think the fog is.

5. Put a small piece of dry ice in the beaker with water. Put a small piece of dry ice in the beaker with oil. Ask students to describe their observations. (Dry ice releases bubbles, evidence that a gas forms. There is a fog over the water but only a very small amount of fog over the oil, evidence that the fog is made up of water droplets. Water has condensed out of the air because it is cooled by the dry ice. Nothing is visible inside the gas bubbles in the oil, indicating that carbon dioxide gas is not visible.)

➡ Provide students with a basic definition of sublimation.

> **Sublimation:** The phase change from a solid to a gas.

Continue the Laboratory Activity

➡ Pass out worksheets to individual students. Ask them to complete Part 1 of the worksheet.

➡ When 10–15 min remain in the class period, ask groups to complete Part 2 of the worksheet—finding the volume of their sublimed gas.

LESSON 8 LAB

It's Sublime
Gas Density

Name _____
Date _____ Period _____

Purpose
To measure density and volume changes associated with phase changes.

Part 1: Observations and Evidence

1. Provide at least three differences you observed between the dry ice and the water ice.

   The water ice melted to a liquid, but the dry ice did not. A fog bubbled up out of the water when the dry ice was put into it, but the water ice simply sat there. Clear bubbles came out of the dry ice when it was put into the oil, and there was no fog.

2. Provide three pieces of evidence that dry ice sublimes.

   The solid disappears slowly. There is no liquid when you heat it. There are bubbles in the oil when the solid disappears.

3. What evidence do you have that the fog is droplets of water and not carbon dioxide?

   Not much fog is visible when the dry ice is placed in oil compared to when it is placed in water.

4. Why does the bag expand?

   Answers may vary.

5. How is the density of the carbon dioxide changing when it sublimes? Explain.

   The density is decreasing, because the same amount of mass is occupying more volume.

6. The mass and volume of two samples of $CO_2(s)$ are measured and graphed. A straight line is drawn through the two data points, beginning at the origin.

   a. What is the density of solid CO_2?

   $$D = \frac{m}{V} = \frac{80 \text{ g}}{50 \text{ mL}} = 1.6 \text{ g/mL}$$

   b. Use the graph for solid CO_2 to show that the value of m/V is the same for a sample with a volume of 40 mL.

   The mass of $CO_2(s)$ with a volume of 40 mL is 63 g, so m/V is 1.6 g/mL.

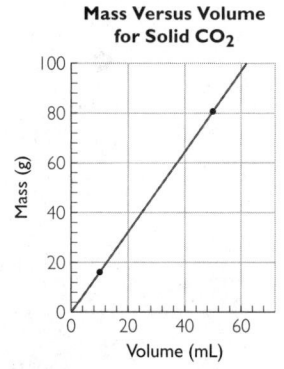

7. Do you predict that the slope of a graph of mass versus volume of gaseous CO_2 will be steeper or less steep than that of the graph of solid CO_2?

   The line will be much less steep, because the gas occupies a larger volume. Thus, the value of m/V will be smaller.

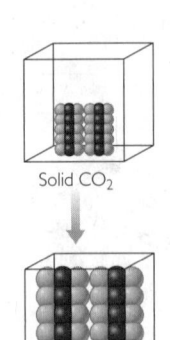

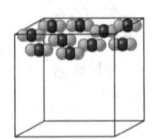

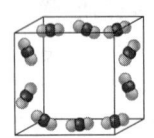

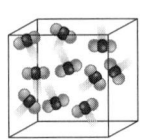

8. These sketches show a model for dry ice and four possible models for what might be happening in your garbage bag as the dry ice sublimes into a gas.

   a. What evidence do you have from your observations of gases that would suggest Models A, B, and C are incorrect?

   You can move through gases. This would suggest that the gas molecules do not occupy much of the space. Also, if they got bigger, it might imply that they have more mass, but the mass does not change. So Model A is incorrect. The garbage bag is pushed out in all directions by the carbon dioxide gas, not just at the top. So Model B is incorrect. If all the molecules were located along the inside surface of the bag, there would be no molecules in the middle. However, we know from experience that when we walk into a room there is air throughout the room. So Model C is incorrect.

   b. How does Model D explain why the bag inflates?

   If the molecules are moving around inside the bag, then each time they collide with the bag, it will get pushed out. That's why the volume increases. Model D is correct.

Part 2: Finish the Laboratory Experiment

Procedure

Once the solid is gone, compress the bag into the measurement box to determine its volume. Measure the height, width, and length in centimeters. Multiply the three measurements to obtain the volume in cubic centimeters. Recall that 1 cm^3 is 1 mL. Record the mass and volume of your group's bag in the table. Find the density of your sample of CO_2.

Mass of CO_2 solid (g)	Volume of CO_2 gas (mL)	Density of CO_2 gas (g/mL)
5.0 g	2500 mL	0.00200 g/mL

Questions

1. How close do you think your answer is to the actual density of $CO_2(g)$? What factors may have affected the accuracy of your measurements?

   Answers will vary. Typically, some CO_2 gas escapes before the bag is closed, so the volume measurement is smaller than it should be, resulting in a higher calculated value for density.

2. Find the volume occupied by your group's carbon dioxide when it was a solid. Use $D = 1.56$ g/mL as the density value for solid CO_2.

   Sample answer: 5.0 g/1.56 g/mL = 3.2 mL.

3. **Making Sense** Why is it easy for a solid object to move through a gas?

   Sample answer: Gases have very low densities. The gas molecules are so far apart and tiny.

Explain and Elaborate (15 minutes)
Discuss the Dry Ice Demonstration
Sample Questions
- What happens when beakers of dry ice and water ice are put on the hot plate?
- Why is a fog not visible when you put the dry ice in oil?
- What do you think the "fog" is that you see?
- Explain the difference between fog and water vapor.

Key Points
Heating water ice and dry ice causes phase changes in both. When ice made from water is heated, it melts into a liquid. If heating continues, the liquid water will become water vapor, a gas. Dry ice is solid carbon dioxide. When dry ice is heated, it becomes a gas, skipping the liquid phase. This is why it is called "dry ice." When a solid is heated to become a liquid, the process is called melting. When a liquid is heated to become a gas, it is called evaporation. When a solid substance becomes a gas without forming a liquid, it is called sublimation.

The "fog" you see around dry ice is actually tiny droplets of liquid water. The fog forms in the vicinity of the dry ice because the temperature is low enough to cause water vapor in the air to become a liquid. Notice that you cannot see the water vapor in the air until it is cold enough to condense. Gaseous carbon dioxide and gaseous water are not visible.

Discuss the Mass and Density of Carbon Dioxide Gas
Sample Questions
- How many times larger is the volume of $CO_2(g)$ than that of $CO_2(s)$? (~800 times larger)
- You determined the mass of the solid CO_2. Is it the same as the mass of the gaseous CO_2 in the bag? Why or why not?
- What density did you calculate for $CO_2(g)$ using your own data?
- Why does the gas feel as if it weighs very little compared to the solid?

Key Points
The space that gaseous CO_2 can occupy is dramatically larger than the space that solid CO_2 occupies. The volume of gaseous carbon dioxide is about 800 times larger than the volume of the solid. The mass of the gas is the same as the mass of the solid, because no molecules are lost or added. Thus, by finding the mass of the solid CO_2 beforehand and measuring the volume of the gaseous CO_2 afterward, it is possible to calculate the density of gaseous CO_2. The actual density of $CO_2(g)$ is 0.0019 g/mL.

When any substance changes into a gas, it expands dramatically. The main substance that condenses and evaporates in large quantities on our planet is water. The large volume change associated with water evaporation and condensation is one key to the dynamics of weather.

> **Evaporation:** The phase change from a liquid to a gas.

Discuss accuracy and possible sources of error. The accepted value for the density of $CO_2(g)$ is 0.0019 g/cm³. Students' measurements usually lead to a higher calculated value. Some CO_2 usually sublimes before the bag is sealed, resulting in a lower volume measurement, which leads to a higher calculated value for density. You can have students calculate their percent error.

$$\% \text{ error} = \frac{|\text{experimental value} - \text{accepted value}|}{\text{accepted value}} \cdot 100$$

Discuss Particle Views of Gases

➡ Draw pictures on the board similar to those on the worksheet.

Sample Questions

- What happens to the solid carbon dioxide molecules when dry ice sublimes?
- How can you explain the large increase in volume when carbon dioxide sublimes?
- It is much easier to compress gases than solids. Is this consistent with the models? Explain.
- Does each model explain why airplanes can fly through clouds? Explain.

Key Points

When solid carbon dioxide sublimes, the individual molecules move farther and farther apart, as shown in Model D. Gas molecules do not individually expand, as shown in Model A, nor do they all float to the top of a container or the surfaces of the container, as shown in Models B and C. Instead, the container inflates because the gas molecules move around. They are relatively far apart from one another, so much of the total volume is empty space.

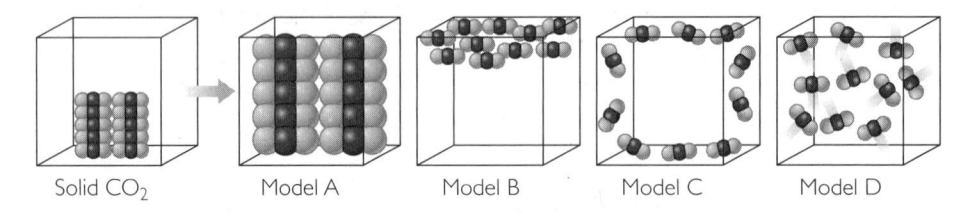

Solid CO_2 Model A   Model B   Model C   Model D

The molecules of a gas are about 1000 times more dispersed (less dense) than the molecules of a solid. In fact, you could say that this is essentially what a gas is—molecules (or atoms) of a substance extremely far apart from one another.

The drawing in Model D is not completely accurate. One improvement that could be made is to place much more distance between the gas molecules in order to reflect a density 800 times lower than that of the solid. However with molecules shown large enough to be seen, this would mean having only one CO_2 molecule in the box, with the next one several feet away.

Wrap-up

Key Question: How do the densities of a solid and a gas compare?

- Sublimation occurs when a substance goes directly from a solid phase to a gas phase.
- When a substance changes phase, its density changes. Individual molecules in a gas are very far apart compared to the molecules in a solid.

- The density of a gas is about 1/1000 the density of the same substance when it is a solid.

Evaluate (5 minutes)

Check-in

A sample of oxygen gas has a mass of 1.43 g and occupies a volume of almost exactly 1000 mL. What is the density of this oxygen gas? Is it more or less dense than carbon dioxide gas?

Answer: Density is equal to mass divided by volume. 1.43 g divided by 1000 mL equals 0.00143 g/mL or g/cm^3. This gas is less dense than carbon dioxide gas: 0.00143 < 0.0019.

Homework

Assign the reading and exercises for Weather Lesson 8 in the student text.

LESSON 9

Air Force
Air Pressure

OVERVIEW

Lesson Type
Demos:
  Whole Class

Key Ideas

Gas pressure can be defined as the force per unit area that results from gas molecules striking the walls of whatever container or object they come in contact with. There is air pressure from the gases that naturally surround us, all the time. This air pressure is called atmospheric pressure.

As a result of this lesson, students will be able to

- describe and define gas pressure
- explain what causes air pressure
- complete simple air pressure calculations

Focus on Understanding

- Students frequently find the idea of the pressure of the air around them baffling, because they cannot feel it. They need to see evidence of its presence, because pressure differentials affect so many situations involving gases. This entire section is devoted to the concept of pressure.

Key Terms

pressure
atmospheric pressure
atmospheres (atm)

What Takes Place

Students view a series of demonstrations involving air pressure. They write down their observations, make drawings to show the various pressures, and discuss with a partner what they think happened. Then each demonstration is discussed as a class. Much of the overall discussion is integrated into the activity. Finally, the concepts of gas pressure, atmosphere, and atmospheric pressure are generally defined.

Note: The composition and properties of the atmosphere are covered more extensively in the student text.

Materials

- student worksheet

Demonstration materials

- several balloons
- 2-liter plastic bottle

74   Living By Chemistry Teacher Guide   Unit 3 Weather

- 1 or 2 empty aluminum soda cans
- hot plate
- tap water
- pair of tongs
- dishpan
- sheet of paper
- clear plastic cup
- large beaker
- rubber air pressure mat
- laminated card to fit over mouth of cup
- large, shallow plastic tub (to catch spills)
- hand vacuum pump and chamber
- fresh marshmallows to be expanded in a vacuum chamber (or small marshmallows to be expanded inside a syringe with a screw-on cap)

Setup

Practice the demos. Instructions for seven demos are included. You can set up to do all or some, depending on the time and equipment available. Consumer hand-pump vacuum chambers for preserving food are available in stores or online.

LESSON 9 GUIDE

Air Force
Air Pressure

Engage (5 minutes)

Key Question: What evidence do we have that gases exert pressure?

Optional: Use a balloon to demonstrate the ChemCatalyst.

> **ChemCatalyst**
>
> If you blow up a balloon and let it go, it flies around the room.
> 1. Why does the gas inside the balloon come out?
> 2. How can you change how fast the balloon moves?
> 3. How does this demonstration provide evidence of air pressure?

Sample Answers: 1. Some students might say that the gas comes out because there is an opening in the container. Others might say it is because there is more pressure in the balloon. 2. If you put less air in the balloon, it will move slower. 3. This demonstration shows that air molecules have mass and can exert pressure, like the wind.

Discuss the ChemCatalyst

Sample Questions

- Why doesn't the balloon stay in one place when you let go of it?
- Why is the balloon moving so fast? How can you change that?
- What do you think air pressure is? What does it have to do with weather?

Explore (20–25 minutes)

Perform the Demonstrations

- You will be conducting a series of demonstrations with some assistance from the class. Students can work individually or in pairs on their worksheets.

- Seven different demonstrations are described. You can complete all or some of them depending on the availability of materials and your time constraints.

- After each demo, ask for possible explanations. Provide time for students to record their observations and answers in the chart on the worksheet. You may want to suggest that students draw diagrams on a separate sheet of paper to explain what they observed for each demo.

- Ask students to consider air pressure inside *and* outside each container. They should use arrows to indicate where there is air pressure. More arrows means greater air pressure.

76 *Living By Chemistry Teacher Guide* Unit 3 Weather

1. BALLOON IN A BOTTLE

Note: For health reasons, you will need one balloon for each student who tries to do the demonstration.

Put an uninflated balloon inside a 2 L plastic bottle. Fold the opening of the balloon back over the mouth of the bottle so it stays in place. Try to blow air into the balloon inside the bottle. Do this as a race, with one student using the balloon inside the bottle setup and another blowing up a balloon in the usual way (no bottle).

Explain: Why was the balloon in the bottle so hard to inflate?

2. SOFT DRINK CAN

Put 5 mL of water in the bottom of an empty aluminum soft drink can. Heat the can on a hot plate until you see steam coming out of the opening. Holding the can with a pair of tongs, quickly invert the can into a dishpan filled halfway with cold water. The can will collapse suddenly and dramatically.

Explain: What causes the can to collapse?

3. SUBMERGED CUP

Fill a large beaker about two-thirds full with water. Crumple a dry piece of paper and squeeze it into the bottom of a plastic cup. Invert the cup, making sure that the paper stays up in the cup. Immerse the cup with the paper completely underwater, holding it as vertical as possible. Take the crumpled paper out of the cup to show that it remained dry.

Explain: Why didn't the paper in the cup get wet?

4. AIR PRESSURE MAT

Place an air pressure mat on a smooth, flat surface. Show that it is easy to pick up the mat if you grab one of the corners and peel it up. Then try to lift the mat by holding onto the hook in the middle. It is impossible to pick it up this way. Allow students to try.

Explain: Why is it so difficult to lift the mat by holding onto the hook?

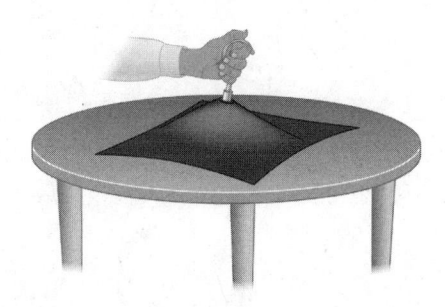

5. CUP AND CARD

Fill a clear plastic cup with water. Place a laminated card over the top of the cup. Hold the card to the mouth of the cup and invert over a plastic tub. You can now let go of the card—it remains suspended, and the water does not spill out.

Explain: Why doesn't the card fall?

6. BALLOON IN A VACUUM

Inflate a balloon to about 2 or 3 in. in diameter. Tie it off. Place the balloon inside a hand-pump vacuum chamber. Put on the lid, and pump to remove the air. This will result in an increase in the size of the balloon. Then allow the air back in to decrease the size of the balloon.

Explain: Why does the balloon increase in size in the vacuum chamber?

7. MARSHMALLOWS

Place several marshmallows in the hand-pump vacuum chamber (or put mini-marshmallows into a large syringe and put the cap on). Pump to remove the air. The marshmallows will puff up and increase in size. Then allow the air back in to decrease the marshmallows' size. The marshmallows will be smaller than they were when you started.

Explain: Why do the marshmallows increase in size inside the vacuum chamber? Why is the final size of the marshmallows smaller than their original size?

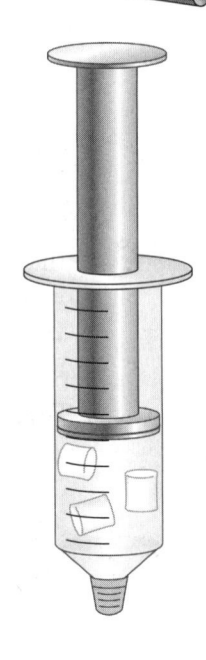

LESSON 9 DEMO

Air Force
Air Pressure

Name _____
Date _____ Period _____

Purpose
To observe and explore situations that involve air pressure.

Directions
Write your observations and explanation for each demonstration.

Demonstration	Observations	Explain what happened
balloon in a bottle	Difficult to blow up balloon in bottle.	No room for more air to go into the bottle.
soft drink can	Can heated with water inside collapses when cooled.	The water changed into a gas, expanding it, then back into a liquid, making it contract.
submerged cup	Paper inside cup stays dry. Air stays inside cup as it is submerged.	The water never touched the paper. The air took up space in the cup.
air pressure mat	It is impossible to lift the mat by the hook in the center.	There's more pressure in the middle of the mat than at the edges.
cup and card	Card doesn't fall, even though there is water pushing on it.	There is more pressure pushing up than down.
balloon in a vacuum	Balloon expands in size as air is removed then shrinks when air is put back.	Removing air from the chamber makes more room for the air in the balloon to expand.
marshmallows	Marshmallows expand when air is removed, then shrink when air is put back in.	There must be air in the marshmallows.

Questions

1. What evidence from the demonstrations shows that air pressure exists?

   Sample answers: The air pressure in the room crushed the can. There is air pressure on the mat we couldn't lift up.

2. **Making Sense** In your own words, describe what you think air pressure is.

   Answers will vary.

228 Unit 3 Weather
Lesson 9 • Worksheet

Living By Chemistry Teaching and Classroom Masters: Units 1–3
© 2010 Key Curriculum Press

Explain and Elaborate (10 minutes)

Discuss Each Demonstration

➡ Go over the demonstrations. Ask students to explain what they think happened in each demonstration.

➡ Ask groups of students to put their drawings on the board, with arrows showing air pressures. You might want to complete this *during* the demonstrations.

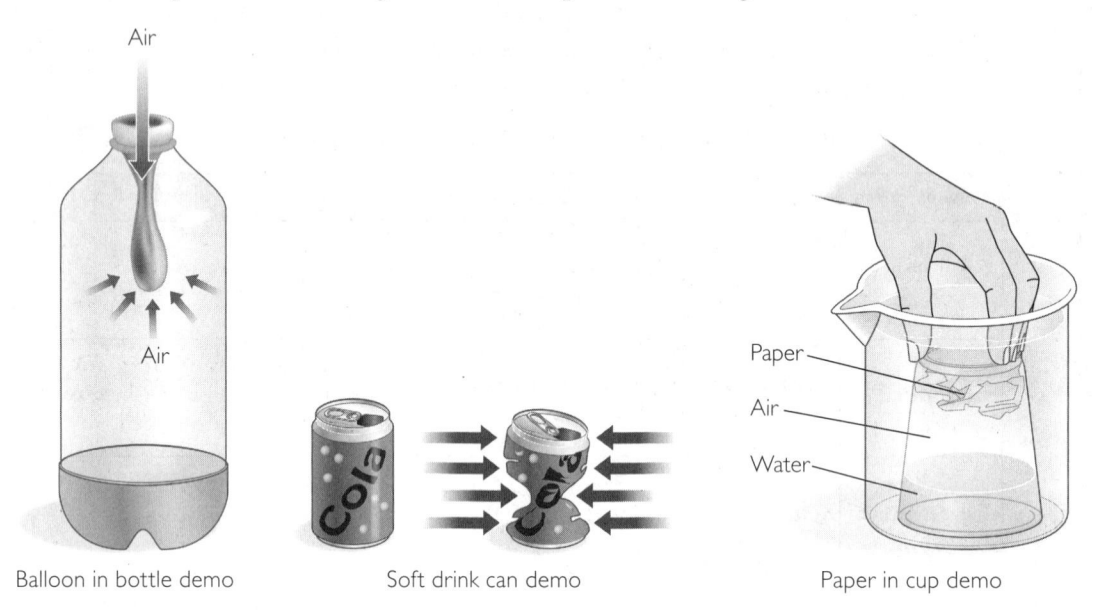

Balloon in bottle demo   Soft drink can demo   Paper in cup demo

Sample Question

- How is air pressure involved in each demonstration?

Key Points

In each demonstration, air is trapped somewhere. In each demonstration, either the pressure of the trapped air is changed or the pressure of the air outside the container with the trapped air is changed.

- *Balloon in a bottle:* The bottle already has air in it. When you try to put air into the balloon inside the bottle, you're pushing against the pressure of the air already in the container.
- *Soft drink can:* Heating the can produces water vapor inside it. When the can is placed upside down in cold water, the opening is sealed off. The water vapor inside the can cools quickly and turns into liquid water. The result is a dramatic decrease in air pressure inside the can. The can collapses because the air pressure on the outside of the can is suddenly so much greater than the pressure on the inside.
- *Submerged cup:* When the cup is inverted in water, there is air inside the cup. This air takes up space. As a result, the water goes only partway up the inside of the cup. The paper stays dry.
- *Air pressure mat:* The mat is held down by atmospheric pressure. Because it is flat against the table, there is no air underneath pushing up on it. (*Note:* Atmospheric pressure is approximately 15 lb/in^2. The mat is about 100 in^2. Thus, the total force pushing down on the mat is about 1500 lb!)

- *Cup and card:* The air pressure pushing up on the card from the atmosphere is larger than the weight of the water and the card pushing down. The result is that the card is held in place.
- *Balloon in a vacuum:* The pressure from the air outside the balloon decreases as the air is pumped out of the chamber. Thus, the balloon increases in size until the air pressure outside the balloon and the air pressure inside the balloon are equal.
- *Marshmallows:* Marshmallows have tiny pockets of trapped air inside them. When the air outside the marshmallows is removed, the air trapped inside the marshmallows expands and the marshmallows puff up. The air pressure in the pockets inside the marshmallows is equalizing with the air pressure outside the marshmallows. Some of the air pockets within the marshmallows will burst, so when air is put back into the chamber the marshmallows actually may be smaller than they were initially.

Define Air Pressure

Sample Questions

- What observations in the demonstrations provide evidence that gases take up space?
- Provide evidence from the demonstrations to support the claim that gases are compressible.
- What observations provide evidence that gases exert pressure?
- What is air pressure?

Key Points

Air pressure is the force per unit area exerted on objects as a result of gas molecules colliding with those objects. When gas molecules are trapped in a container, those gas molecules collide with and push on the walls of the container. Thus, gas pressure often is expressed in pounds per square inch, or lb/in^2. Even though gas molecules are very far apart from one another compared to the molecules in a solid or a liquid, they can account for a great deal of force.

> **Pressure:** Force applied over a specific area. Force per unit area. Gas pressure is caused by gas molecules striking objects or the walls of a container.

In examples like the ones seen today, there are generally two types of air pressure to consider: air trapped inside a container, and air from the atmosphere outside a container. All gases exert gas pressure, no matter what their identity is. It is important to note that all the things we learn about air pressure apply to other gases as well, such as helium, oxygen, and fluorine.

The mixture of gases that surrounds you at all times is called the atmosphere. We commonly refer to it simply as "air." There is constant pressure on you from this air. While you might not feel the air pushing against you, the gases in the air exert a significant amount of pressure, called atmospheric pressure. You are so used to atmospheric pressure that you usually forget it is there.

> **Atmospheric pressure:** Air pressure that is always present on Earth as a result of air molecules colliding with the surfaces of objects on the planet. At sea level and 25 °C, there is 14.7 lb/in^2 of air pressure from the air around us. This is referred to as one atmosphere of pressure, or 1 atm.

Wrap-up

Key Question: What evidence do we have that gases exert pressure?

- Gas pressure is defined as the force per unit area caused by the molecules of a gas pushing on objects or on the walls of its container.
- Atmospheric pressure is the pressure of the air around us.
- Many generalizations you make about air can also be made about any gas, whether helium, argon, or any other.

Evaluate (5 minutes)

> **Check-in**
>
> Give evidence that gas molecules exert pressure on the walls of whatever container they are in.

Sample Answer: The fact that a balloon is stretched out tight when it is blown up is one example of the air pressure exerted by gas molecules.

Homework

Assign the reading and exercises for Weather Lesson 9 in the student text.

LESSON 10 OVERVIEW

Feeling Under Pressure
Boyle's Law

Lesson Type
Lab:
  Pairs
Demo:
  Whole Class

Key Ideas

When an amount of gas is compressed into a smaller volume, its pressure increases. The pressure of a gas is inversely proportional to the volume the gas occupies (if the temperature and amount of gas do not change). Thus, when one variable gets larger, the other variable must get smaller, and vice versa. Mathematically, the relationship between pressure and volume is expressed as $P_1V_1 = P_2V_2 = k$, or $P = k/V$. This mathematical relationship is known as Boyle's law.

As a result of this lesson, students will be able to

- explain the relationship between gas pressure and gas volume
- define an inversely proportional relationship
- state Boyle's law

Focus on Understanding

- Among the weather variables, this is the only inversely proportional relationship students will encounter. Thus, it may give them some difficulty initially.
- So far we have avoided using the term constant to mean fixed or unchanging in order to prevent confusion with the proportionality constant. However, in this lesson, you may want to point out this meaning of the word to students.

Key Terms

inverse proportion
Boyle's law

What Takes Place

SciLINKS
Topic: Gas Pressure
Visit: www.SciLinks.org
Web code: KEY-310

Students explore the relationship between gas pressure and gas volume experientially using large syringes. Students work in pairs for the first portion of the lab before gathering in larger groups for Part 2. The larger groups gather pressure and volume data using a syringe with air trapped inside the chamber. A bathroom scale measures the force required to change the volume inside the syringe. (Instead of the scale and syringe, you can use a platform apparatus like the one shown—sometimes called an elasticity of gases apparatus—and have students place a book or other objects with known mass on top.)

Students graph the data and are guided to recognize that the relationship between pressure and volume is an inverse relationship.

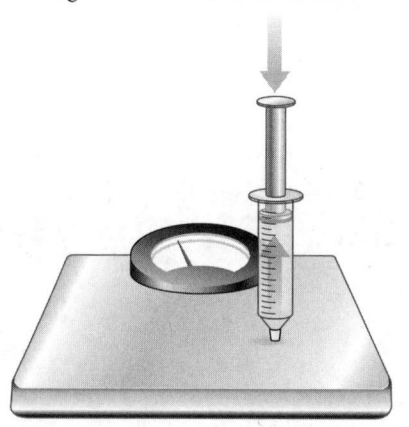

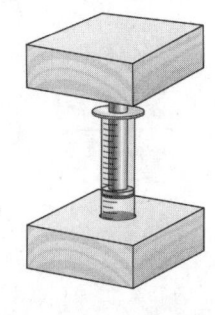

Section II Lesson 10 Feeling Under Pressure

Materials

- student worksheet
- empty plastic water bottle with cork (you might want to put a streamer on the cork for visibility) (optional for ChemCatalyst)
- Math Card—Boyle's law (optional)

Per pair

- 50 mL plastic syringe (without cap)

For demo (or per group of 4 or 8)

- 50 mL plastic syringe (with screw-on cap or cap securely glued onto the tip)
- bathroom scale

Setup

You will need two types of syringes for this lab: uncapped syringes for Part 1 of the lab and capped syringes for Part 2. It is important to check the capped syringes to make sure the caps are tightly screwed on or firmly secured with superglue. Instruct students to wear safety glasses and to *always point the syringe with cap down as they push on it.* The caps can fly off at very high pressures. The number of groups you have for Part 2 will depend on the number of bathroom scales you have. Thus, you could do Part 2 in larger groups or as a student-assisted demo.

LESSON 10 GUIDE

Feeling Under Pressure
Boyle's Law

Engage (5 minutes)

Key Question: How does gas volume affect gas pressure?

> **ChemCatalyst**
>
> An empty plastic water bottle has a cork fitted into the opening.
>
> 1. Predict what would happen if you stepped on the plastic bottle.
> 2. Explain your answer in terms of pressure and volume.

Sample Answers: 1. The cork should fly out of it. 2. The bottle contains trapped air. When you step on the bottle, you decrease the amount of space this air has to fit into causing air pressure to increase inside the bottle. The cork flies out because it is the weakest place in the system.

Discuss the ChemCatalyst

➡ You might want to demonstrate with a bottle and cork. Be sure to point the cork away from anyone and from anything breakable.

Sample Questions

- Is the bottle truly empty? Explain.
- What is happening in terms of pressure and volume?
- If you squeeze some gas into a smaller volume, what happens to the pressure inside the container?

Explore (20 minutes)

Introduce the Lab

➡ Discuss limiting the number of variables. Write the terms *temperature, volume, pressure,* and *number of molecules* on the board. In this lab, the number of molecules doesn't change and the gas is assumed to remain at room temperature. Today's class focuses on pressure and volume and how these two variables interact.

Guide the Lab

➡ Students can work in pairs on Part 1. Depending on the number of bathroom scales you have, you can do Part 2 as a whole-class demo or in groups.

➡ Students are asked to estimate the cross-sectional area inside the syringe. They might need help visualizing or calculating this area. It probably will be about 1–2 in^2, depending on the syringe.

Section II   Lesson 10 Feeling Under Pressure   85

LESSON 10 LAB

Feeling Under Pressure
Boyle's Law

Name _____
Date _____ Period _____

Purpose
To observe and quantify the relationship between gas pressure and gas volume.

Part 1: Observing Pressure
Materials
- 50 mL plastic syringe

Procedure
1. Start the plunger at 50 mL. Cover the tip of the syringe with your fingertip. Be sure to make a good seal.
2. Push the plunger until it reads 40 mL.
3. Continue to cover the tip with your finger. Apply more pressure until the inside volume reads 30 mL, then 20 mL, and so on.

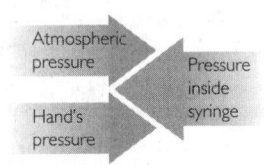

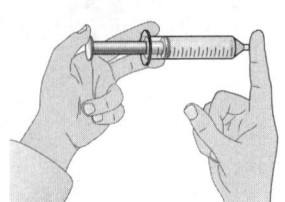

Analysis
1. What did you experience when you pushed in the plunger from 40 mL to 30 mL, and then from 30 mL to 20 mL?

   It is harder to go from 30 mL to 20 mL than from 40 mL to 30 mL.

2. Are you able to push the plunger all the way in? Explain why or why not.

   The plunger cannot be pushed all the way in because the air inside the syringe pushes back.

3. Explain why the number of air molecules in the syringe doesn't change but the volume does.

   The number of air molecules stays the same because the air cannot escape from the inside with a finger sealing the tip. The air molecules are compressed into a smaller space, so the volume changes.

Part 2: Weight Versus Volume Data
Materials
- 50 mL plastic syringe with cap screwed on tight
- bathroom scale

Living By Chemistry Teaching and Classroom Masters: Units 1–3
© 2010 Key Curriculum Press

Unit 3 Weather 229
Lesson 10 • Worksheet

Procedure

> **Safety Instructions**
> ⚠ The cap on the tip of the syringe should always be pointed down, away from eyes. Wear safety glasses.

1. Start with the syringe at 50 mL. Make sure the cap is on tight.
2. Hold the syringe vertically with the tip on top of a bathroom scale.
3. Follow these steps:
    - One person should depress the plunger by a few milliliters.
    - A second person should read the exact volume.
    - A third person should read the number of pounds that is exerted on the bathroom scale.
    - Everyone should record the volume and weight data in a table like the one shown.
    - Repeat these steps for at least five different volumes. Depress the plunger a little more each time. Be sure to include a reading where you depress the plunger as far as you can.

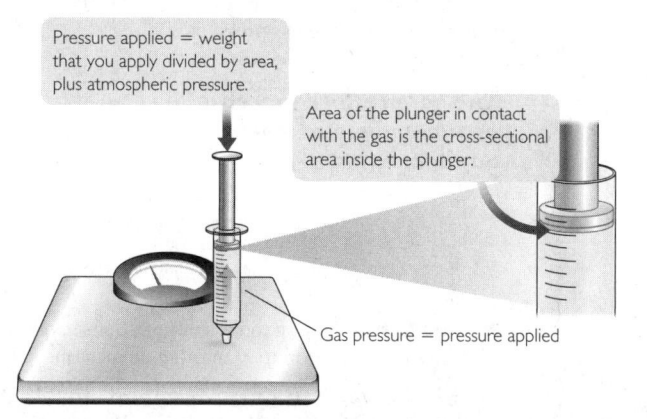

Pressure applied = weight that you apply divided by area, plus atmospheric pressure.

Area of the plunger in contact with the gas is the cross-sectional area inside the plunger.

Gas pressure = pressure applied

Data

Trial	Volume (mL)	Weight, or force, you apply (lb)	Pressure you apply (lb/in^2)	Atmospheric pressure (lb/in^2)	Total pressure (lb/in^2)
1	50 mL	0 lb	0 lb/in^2	14.7 lb/in^2	15 lb/in^2
2	40 mL	3.5 lb	4.4 lb/in^2	14.7 lb/in^2	19 lb/in^2
3	30 mL	8.0 lb	10 lb/in^2	14.7 lb/in^2	25 lb/in^2
4	20 mL	18 lb	23 lb/in^2	14.7 lb/in^2	38 lb/in^2
5	15 mL	28 lb	35 lb/in^2	14.7 lb/in^2	50 lb/in^2
6	10 mL	48 lb	60 lb/in^2	14.7 lb/in^2	75 lb/in^2

Actual data will vary.

Analysis

1. The number of pounds on the bathroom scale is related to the pressure inside the syringe. Explain why.

   The reading on the scale is directly proportional to how much force you exert. The bathroom scale roughly measures the amount of force you are using to push down.

2. Estimate the cross-sectional area inside the syringe in square inches.

   Cross-sectional area inside the syringe: _____ **0.80 in^2** _____.

   Calculate the pressure you applied in pounds per square inch. Enter these values in the "pressure you apply" column.

3. The atmosphere is also applying a pressure on the gas in the syringe. This pressure is equal to 14.7 lb/in^2. Add 14.7 lb/in^2 to the pressure you apply to obtain the total pressure on the gas. Enter these values in the last column.

4. Plot the total pressure in pounds per square inch versus volume on the graph and connect the points with a curve.

5. What happens to the gas pressure as the volume of the gas decreases?

   The gas pressure increases as the volume of the gas decreases.

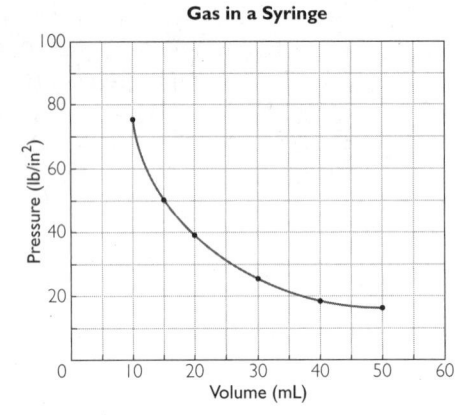

   Gas in a Syringe

6. **Making Sense** Using today's observations, explain how the pressure and volume of a gas change in relation to each other as each one increases or decreases.

   When the volume of a sample of gas decreases, the pressure exerted by that gas increases. When the volume of a gas sample increases, the pressure exerted by that gas decreases.

7. **If You Finish Early** Create a graph of P versus $1/V$. What is the outcome? What does the graph tell you about the relationship between gas pressure and volume?

   The graph is a straight line going toward the origin. This means the relationship between gas pressure and volume is inversely proportional.

Explain and Elaborate (15 minutes)

Discuss Air Pressure as It Relates to the Activity

Sample Questions

- What did you observe when you explored the syringe with no cap on it?
- Does the amount of air change when you push in the plunger?
- Can you decrease the volume of the syringe to zero? Why or why not?
- Why does it get harder and harder to depress the plunger as the volume gets smaller and smaller?

Key Points

In the syringe and scale experiment, you put weight on the gas in the syringe, which you measured in pounds on the scale. Pounds can be converted into pressure by dividing by the area on which you are pushing. In this case, the area is the cross-sectional area of the chamber in the syringe. In today's class you used units of pounds per square inch, or lb/in^2. It is necessary to add 14.7 lb/in^2 to all the measurements to account for the force of the air pressure from the atmosphere.

When the volume of a gas is decreased, its pressure goes up. This is because the same number of molecules must occupy a smaller space. A smaller space containing the same number of molecules means more collisions and therefore higher pressure. Conversely, when the volume of a gas is increased, its pressure goes down. This is because the same number of gas molecules must occupy a larger space. There are fewer collisions, and therefore the gas molecules exert a lower pressure.

Process the Graph of Gas Pressure Versus Gas Volume

➡ Draw *x*- and *y*-axes on the board with Volume on the *x*-axis and Pressure on the *y*-axis. Have students help you create a simple graph using the data from the syringe and scale experiment.

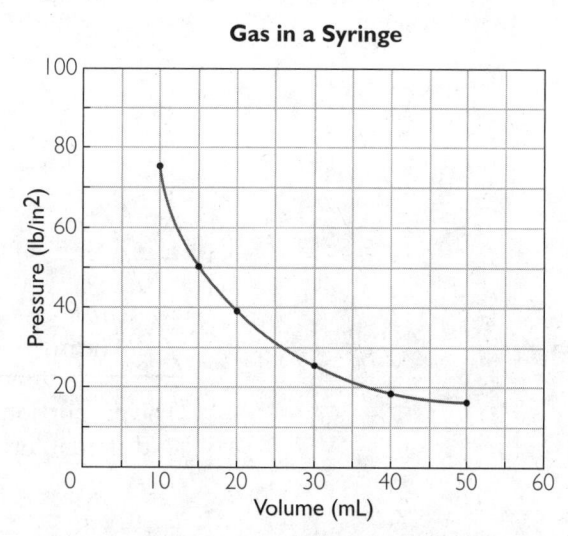

Gas in a Syringe

(Graph showing Pressure (lb/in²) on y-axis from 0 to 100, Volume (mL) on x-axis from 0 to 60, with a decreasing curve starting near (10, 75) and leveling off around (50, 16).)

Sample Questions

- What does the graph show about the relationship between pressure and volume?
- Why isn't the graph a straight line?
- Does the curve ever touch the axis? Explain.
- Does the graph of pressure versus volume describe a proportional relationship? Why or why not?

Key Points

The relationship between gas pressure and gas volume is called an inverse proportion. This means that when one variable increases, the other decreases, and vice versa. The graph of gas pressure versus gas volume is a curve, not a straight line going through the origin. In contrast, with the rain gauge and with Charles's law the variables you encountered are directly proportional, where both variables increase together.

> **Inverse proportion:** Two variables are inversely proportional to each other if one variable increases when the other decreases.

The curve on the graph is nearly vertical for small volumes and nearly horizontal for large volumes. For small volumes, the gas pressure in the syringe is extremely high. After a certain point it is difficult to get the volume any smaller, no matter how hard you push. In other words, you hit a limit as to how small the gas volume can be. At large volumes, the pressure is extremely low in the syringe. Once the volume is large, the pressure doesn't change much. Between the two extremes, changes in volume and pressure affect each other more.

Introduce Boyle's Law

➡ You might want to complete a graph of the data for P versus $1/V$ on the board.

➡ Hand out the Math Card for Boyle's law to students (optional).

Boyle's Law

$$k = P \cdot V$$

k (atm · L)

P (atm) $\quad$ V (L)

The proportionality constant, k, is different for each gas sample.

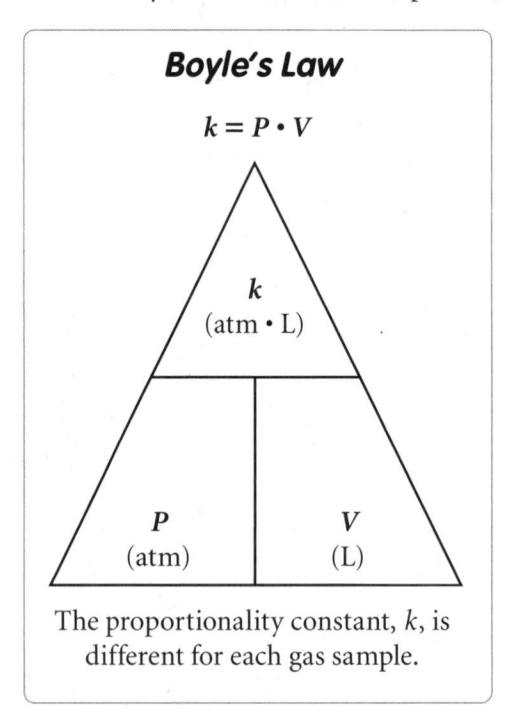

Sample Questions

- What happens if you create a graph of P versus $1/V$?
- What is the approximate value of P times V for your data? (It should be somewhere around 600–700.)
- What would happen to your data if you heated the air in the syringe and then measured pressure and volume? (k is different at a different T.)

Key Points

The mathematical relationship between pressure and volume is described by the equation

$$PV = k \quad \text{or} \quad P = k/V$$

This relationship is known as Boyle's law, after Robert Boyle, the British scientist who discovered the relationship. If you were able to make very accurate measurements in your syringe and scale experiment, the proportionality constant, k, would be the same number for every value of P times V. You can have students calculate the product PV for their data. The products will be nearly identical, but not exactly, due to experimental error.

A graph of P versus $1/V$ is a straight line going through the origin. This means that there is an inversely proportional relationship between P and V. It is important to emphasize that the proportionality constant, k, is a different number if the amount of gas or the temperature is changed. Thus, for each new situation, k has a different value, equal to PV.

> **Boyle's law:** The pressure of a given amount of gas is inversely proportional to its volume, if the temperature and amount of gas are not changed. The relationship between pressure and volume can be expressed as $PV = k$, or $P = k(1/V)$, where k is the proportionality constant.

Boyle's law enables you to solve problems involving gas pressure and gas volume.

Wrap-up

Key Question: How does gas volume affect gas pressure?

- Gas pressure is inversely related to its volume. When the volume of a gas decreases, its pressure increases, and vice versa.
- The mathematical relationship between gas pressure and volume can be described by the formula $PV = k$ or $P = k/V$, provided the temperature and amount of gas are kept the same. This relationship is known as Boyle's law.

Evaluate (5 minutes)

> **Check-in**
>
> A balloon full of gas occupies 7.5 L, and the pressure on the outside of the balloon is 1.0 atm. What do you predict will happen to the pressure inside the balloon if the balloon is placed underwater to a depth at which its new volume is 2.5 L?

Answer: The pressure will increase, because the volume decreases.

Homework

Assign the reading and exercises for Weather Lesson 10 in the student text.

LESSON 11

Egg in a Bottle
Gay-Lussac's Law

OVERVIEW

Lesson Type
Classwork:
 Individuals
Demo:
 Individuals

Key Ideas

Changing the temperature of a gas will change its pressure if the volume is kept the same. The pressure of a gas is proportional to the temperature of a gas if the volume and amount of gas do not change and if the temperature of the gas is expressed in kelvin. Mathematically, the relationship between pressure and temperature is $P = kT$, where k is the proportionality constant. This mathematical relationship was first investigated by Guillaume Amontons in the late 1600s but is often called Gay-Lussac's law. Boyle's law and Gay-Lussac's law are two gas laws that apply to situations in which the pressure changes in response to changes in either volume or temperature. The type of container a gas occupies—rigid or flexible—affects which of these gas laws applies.

As a result of this lesson, students will be able to

- describe the qualitative and quantitative relationships between the pressure and temperature of a gas
- explain how flexible and rigid containers affect the pressure, volume, and temperature of a gas sample
- complete gas law problems involving changes in pressure

Focus on Understanding

- Flexible and rigid containers present some conceptual challenges. For a flexible container, the pressure differences both outside and inside the container must be considered.
- Students often forget about the effects of atmospheric pressure.
- The pressure change in a rigid container can be observed only with a pressure gauge or if the container bursts.
- There are a number of mathematical approaches to solving gas law problems, including using proportional reasoning and dimensional analysis. Students will use whichever method works best for them based on the math strategies they have already learned. Our approach presents the gas laws in such a way as to best assist students with lower math abilities while not getting in the way of students with higher abilities.

Key Term

Gay-Lussac's law

What Takes Place

Students observe a demonstration of Gay-Lussac's law. A hard-boiled egg is forced into an Erlenmeyer flask by cooling the air inside to alter the air pressure

Living By Chemistry Teacher Guide Unit 3 Weather

inside the flask. Students complete a worksheet that allows them to gain some proficiency at completing problems involving Gay-Lussac's law and Boyle's law to determine pressure. Students consider the effects of flexible and rigid containers on the pressure, volume, and temperature of a gas sample. Additional worked examples reinforce the gas laws.

Note: This lesson can be completed in a block period or two standard class periods.

Materials

- student worksheet
- Math Cards—Gay-Lussac's law (optional)

Demonstration materials

- hard-boiled egg, shelled
- 1000 mL Erlenmeyer flask
- oven mitt
- hot plate
- Bunsen burner or blow dryer (optional)

Setup

Prepare hard-boiled eggs the night before class. You need only one egg per class if all goes well, but it is useful to have a few extra in case one breaks. Set up the flask with a small amount of water inside and the egg on top of the flask so students will see the setup when they answer the ChemCatalyst.

LESSON 11 GUIDE

Egg in a Bottle
Gay-Lussac's Law

Engage (5 minutes)

Key Question: How does gas pressure change in flexible and in rigid containers?

→ Assemble the egg and bottle demonstration at the beginning of class so students can look at the setup and answer the ChemCatalyst.

> **ChemCatalyst**
>
> Examine the egg and bottle setup. How could you use gas pressure to get the egg into the bottle? What variables would you change: pressure, volume, and/or temperature?

Sample Answer: Some students might suggest increasing air pressure outside to push on the egg, while others will suggest that the air pressure inside the bottle be reduced. Some might suggest warming the outside air and cooling the air inside the flask. Others will suggest increasing the pressure outside the flask and reducing it inside.

Discuss the ChemCatalyst

→ Discuss ideas about how to get the egg into the bottle.

Sample Questions

- How would you get the egg into the bottle using gas pressure?
- What happens to the gas inside the bottle when it is cooled?
- What force might push the egg into the bottle?

Explore (50 minutes)

Complete the Egg in a Bottle Demonstration

→ Demonstrate how differences in gas pressure can be used to force an egg into a bottle.

1. With a small amount of water in the bottom of the flask, heat the flask on a preheated hot plate for several minutes. (There should be enough water so it does not all boil away.) Remove the flask from the hot plate with an oven mitt when you see steam coming out of the flask.

2. Place the hard-boiled egg on the opening of the flask so it makes a seal. Observe what happens as the air inside the flask cools (the egg gets pushed into the flask).

3. You can speed up the process by placing the flask in cool water or ice water.

4. Ask students to explain what they think happened.

5. Ask students how they could use gas pressure to get the egg back out of the flask. Listen to their suggestions. Then turn the flask upside down so that the egg falls into the opening. Hold the flask so that it is tipped sideways, and reheat the flask on the bottom until the egg is pushed back out of the flask. A Bunsen burner works best for this last procedure, but you can use a hot plate or a blow dryer. Be sure to use an oven mitt when holding the flask.

Introduce the Classwork

→ Ask students to work individually on the worksheets. Tell them they will be exploring a number of situations in which the pressure changes in response to either a change in volume or a change in temperature.

LESSON 11 CLASSWORK

Egg in a Bottle
Gay-Lussac's Law

Name _____
Date _____ Period _____

Purpose
To examine how gas pressure changes in flexible and rigid containers.

Part 1: Glass Bottle (Rigid Container)

The air trapped inside a 240 mL glass bottle has a pressure of 1.0 atm and a temperature of 25.0 °C. You put the glass bottle into a freezer. After several hours, the air trapped inside the bottle has a temperature of −35.0 °C and a pressure of 0.80 atm.

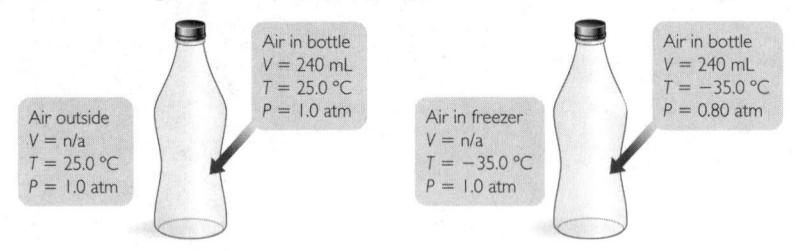

Air outside
V = n/a
T = 25.0 °C
P = 1.0 atm

Air in bottle
V = 240 mL
T = 25.0 °C
P = 1.0 atm

Air in freezer
V = n/a
T = −35.0 °C
P = 1.0 atm

Air in bottle
V = 240 mL
T = −35.0 °C
P = 0.80 atm

1. When the glass bottle is put into the freezer, how does the air trapped inside the bottle change? How is it the same?

   Gas temperature and pressure change. The volume of the trapped air does not change, because the bottle is rigid.

2. Determine the value of k for the air trapped inside the glass bottle before and after cooling to show that P equals kT.

   First, convert the temperature to kelvins by adding 273 K:
   $T_1 = 25.0 \,°C + 273 = 298 \,K \qquad T_2 = -35.0 \,°C + 273 = 238 \,K$
   Then calculate k:
   $$k = \frac{P_1}{T_1} = \frac{1.0 \text{ atm}}{298 \text{ K}} = 0.0034 \text{ atm/K}$$
   $$k = \frac{P_2}{T_2} = \frac{0.80 \text{ atm}}{238 \text{ K}} = 0.0034 \text{ atm/K}$$

3. The table shows data for the glass bottle in four locations. The atmospheric pressure stays unchanged at 1.0 atm, but the temperature is different at each location.

   a. Complete the table.

Air outside the bottle			Air inside the bottle			
V (mL)	T (K)	P (atm)	V (mL)	T (K)	P (atm)	P/T
N/A	200 K	1.0 atm	240 mL	200 K	0.67 atm	0.0034 atm/K

Air outside the bottle			Air inside the bottle			
V (mL)	T (K)	P (atm)	V (mL)	T (K)	P (atm)	P/T
N/A	238 K	1.0 atm	240 mL	238 K	0.80 atm	0.0034 atm/K
N/A	298 K	1.0 atm	240 mL	298 K	1.0 atm	0.0034 atm/K
N/A	400 K	1.0 atm	240 mL	400 K	1.34 atm	0.0034 atm/K

b. Plot pressure versus temperature for the air inside the glass bottle at each location. Draw the best-fit line through the data points to determine the pressure inside the glass bottle if the temperature is 350 K.

1.2 atm

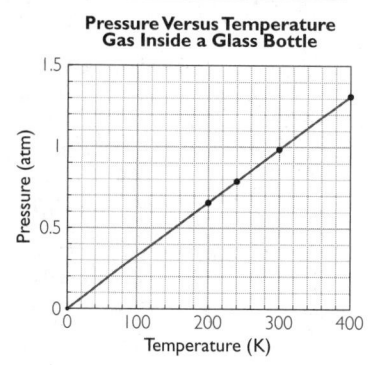

Pressure Versus Temperature Gas Inside a Glass Bottle

4. A sample of chilled air from a freezer is sealed up inside a glass bottle with a volume of 240 mL. This bottle is then allowed to warm to room temperature. What is the air pressure inside the bottle at 25 °C? Show your work.

Air in the bottle in the freezer

P_1	1.0 atm
T_1	−35 °C
V_1	240 mL

Air in the bottle in the room

P_2	—?—
T_2	25 °C
V_2	240 mL

$$k = \frac{P_1}{T_1} = \frac{1.0 \text{ atm}}{238 \text{ K}} = 0.0042 \text{ atm/K}$$
$$P_2 = kT_2 = (0.0042 \text{ atm/K})(298 \text{ K}) = 1.3 \text{ atm}$$

Part 2: Car Air Bag (Flexible Container)

Identical air bags inflate in two different cars. One car is at sea level, and the second car is in the mountains. The temperature, pressure, and volume of the air outside the air bag and of the gas inside the air bag are given in the table.

	Air at sea level	Gas in airbag at sea level	Air on mountain	Gas in airbag on mountain
V	–	60.0 L	–	86.0 L
T	25.0 °C	25.0 °C	25.0 °C	25.0 °C
P	1.0 atm	1.0 atm	0.70 atm	0.70 atm

1. Consider the gas trapped inside the air bag. How do the volume, pressure, and temperature of the gas change as you go from sea level to the mountaintop?

   Gas volume and pressure change. The temperature of the air outside does not change, so the temperature of the gas inside the air bag is the same in both locations.

2. Why is the volume of the air bag different in the two locations?

   The pressure of the air outside the air bag changes. The volume of the air bag changes so that the pressure of the gas inside the air bag matches the pressure of the air outside.

3. Show that the pressure of the gas in the air bag is inversely proportional to the volume of the gas in the air bag (Boyle's law).

   $k = PV$
   Sea level: $1.0 \text{ atm} \cdot 60 \text{ L} = 60 \text{ L} \cdot \text{atm}$
   Mountaintop: $0.70 \text{ atm} \cdot 86 \text{ L} = 60 \text{ L} \cdot \text{atm}$

4. A different car has an air bag that inflates to 60 L on the mountaintop, where the air pressure is 0.70 atm and the temperature is 25 °C. What volume will this air bag have at sea level?

Mountaintop	
P_1	0.70 atm
T_1	25 °C
V_1	60.0 L

Sea level	
P_2	1.0 atm
T_2	25 °C
V_2	—?—

   $k = P_1 V_1 = 0.70 \text{ atm} \cdot 60.0 \text{ L} = 42 \text{ L} \cdot \text{atm}$
   $V_2 = \dfrac{k}{P_2} = \dfrac{42 \text{ L} \cdot \text{atm}}{1.0 \text{ atm}} = 42 \text{ L}$

5. **Making Sense** Compare a rigid container, such as a glass bottle, with a flexible container, such as an air bag. Describe how the type of container affects how the pressure of the gas inside the container can vary.

   For a rigid container, the pressure of the trapped gas does not always match the pressure of the air outside. The volume cannot change, so the only way to change the pressure of the trapped gas is to change the temperature.

   For a flexible container, the pressure of the trapped gas will change to match the pressure of the air outside. You can change the pressure of the trapped gas by changing the temperature, or by changing the air pressure on the outside.

Explain and Elaborate (25 minutes)

Briefly Discuss the Demonstration

Sample Questions

- Explain what you think caused the egg to go into the flask.
- The egg was pushed, not sucked, into the flask. Explain what this means.

Key Point

As the gas inside the flask cools, the pressure of the gas inside the flask decreases. Because the air pressure decreases inside the flask, the force on the egg exerted by the air *outside* the flask is greater than the force exerted on the egg by the air *inside* the flask. When the difference becomes great enough, the egg is pushed into the flask. The change in gas pressure happens both because gas pressure is proportional to temperature and because some of the water vapor in the flask becomes a liquid. Notice that the egg is not "sucked" into the flask. It is pushed into the flask by the higher air pressure outside the flask. Gaseous molecules cannot "suck."

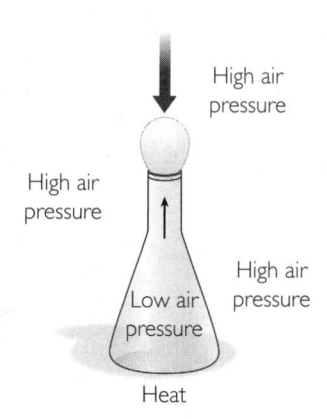

Introduce Gay-Lussac's Law

Sample Questions

- Describe a situation in which you would use Gay-Lussac's law.
- Why is it dangerous to heat a sealed container?
- How does the type of gas container (rigid or flexible) relate to this gas law?

Key Point

Gas pressure is proportional to temperature if the volume and amount of gas are not changed and if the temperature is expressed in kelvins. This relationship is known as Gay-Lussac's law. It is named after French scientist Joseph Louis Gay-Lussac, who described the relationship in 1802. It is expressed as $P = kT$. The proportionality constant, k, describes how much the pressure increases per degree of temperature in kelvins. The value of k for a set of conditions is P/T. Gay-Lussac's law applies only to rigid containers, for which the volume does not change.

> **Gay-Lussac's law:** The pressure of a given amount of gas is directly proportional to temperature if the gas volume and amount of gas do not change and if the temperature is expressed in kelvins.

Compare Gases in Rigid and Flexible Containers

➡ Tell students to take out the Math Cards for Charles's law, Boyle's law, and Gay-Lussac's law (optional).

Gay-Lussac's Law

$$k = \frac{P}{T}$$

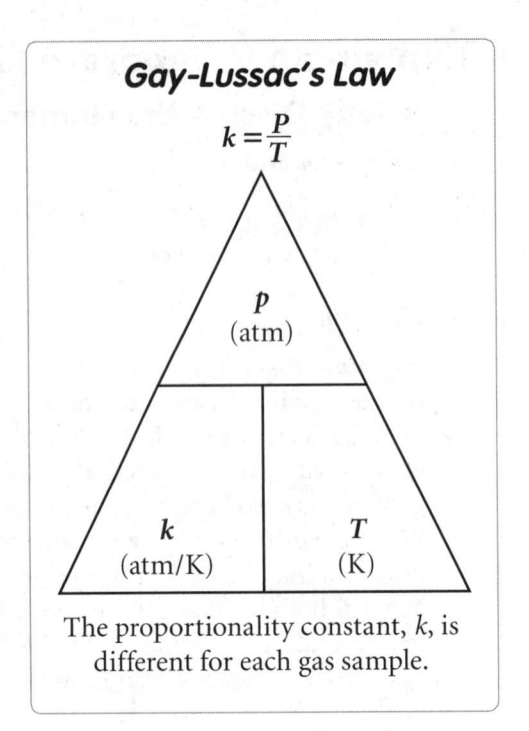

```
        P
      (atm)

   k           T
 (atm/K)      (K)
```

The proportionality constant, k, is different for each gas sample.

Sample Questions
- Give examples of flexible containers. (balloon, piston and cylinder, syringe, bag)
- Give examples of rigid containers. (bottle, can, gas tank, pressure cooker)
- How does the type of gas container (rigid or flexible) relate to Charles's law, Boyle's law, and Gay-Lussac's law?

Key Points

The three gas laws help predict gas temperature, pressure, and volume when two of these variables change and the third remains fixed. There are two types of containers for the gas: one that can vary in volume, such as a balloon or a syringe, and one with a fixed volume, such as a sealed glass bottle.

- *Flexible container:* When the gas is in a *flexible* container, changing the temperature or the pressure causes the volume to change. Charles's law applies when volume and temperature vary but pressure does not change. Boyle's law applies when volume and pressure vary but temperature does not change.
- *Rigid container:* When a gas is in a *rigid* container, increasing the temperature causes the pressure to increase. Gay-Lussac's law applies when the pressure and temperature vary but volume does not change.

It is important to notice that in all these cases gas cannot enter or escape the containers. The amount of gas is fixed. We will explore the effects of changing the number of gas molecules in the next section.

Solve Sample Gas Law Problems

Keys to solving gas law problems
1. Identify which variable is *not* changing: P, V, or T.
2. Identify the two variables that *are* changing: P and V, P and T, or V and T.
3. Identify the gas law formula that should be used to solve the problem.
4. Determine the proportionality constant k from P_1, V_1, or T_1.

5. Use k to find P_2, V_2, or T_2.

6. Remember that all temperatures must be in kelvins.

Example 1

Imagine that you have a syringe with a plunger that slides easily. There is air trapped inside. The gas pressure is 1.0 atm, the gas temperature is 20 °C, and the volume of the gas is 52 mL. You place the syringe in the freezer when the gas pressure is 1.0 atm. The volume of the air trapped in the syringe changes to 45 mL, and the pressure is still 1.0 atm. What is the temperature of the air in the freezer?

Solution

Note that the pressure of the air in the syringe is the same as the pressure of the air outside, because the container is flexible.

P_1	1.0 atm
T_1	20 °C
V_1	52 mL

P_2	1.0 atm
T_2	—?—
V_2	45 mL

You need to use the relationship between V and T (Charles's law).

Always convert the temperature to kelvins:

$$T_1 = 20\ °C + 273 = 293\ K$$

Then calculate k. Use the volume, V_1, and the temperature, T_1, of the air trapped in the syringe:

$$k = \frac{V_1}{T_1}$$
$$= \frac{52\ mL}{293\ K}$$
$$= 0.18\ mL/K$$

Next, use k to determine T_2, the temperature inside the freezer:

$$T_2 = \frac{V_2}{k} = \frac{45\ mL}{0.18\ mL/K} = 250\ K = -23\ °C$$

The temperature of the air in the syringe decreased. This is why the volume decreased.

Example 2

A gas is sealed in a 35 mL metal can at a pressure of 1.0 atm and a temperature of 25 °C. The can is rated for a maximum pressure of 2.0 atm. You heat the can until the pressure relief valve bursts. What is the temperature of the gas when the pressure relief valve bursts?

P_1	1.0 atm
T_1	25 °C
V_1	35 mL

P_2	2.0 atm
T_2	—?—
V_2	35 mL

Solution

The volume does not change in this example. You need to use the relationship between P and T (Gay-Lussac's law).

Convert the temperature to kelvins:

$$T_1 = 25\,°C + 273$$
$$= 298\,K$$

Calculate k. Use the initial pressure, P_1, and the initial temperature, T_1, of the air trapped in the can:

$$k = \frac{P_1}{T_1} = \frac{1.0\,\text{atm}}{298\,K} = 0.0034\,\text{atm/K}$$

Next, use k to determine T_2, the temperature at which the can bursts:

$$T_2 = \frac{P_2}{k}$$
$$= \frac{2.0\,\text{atm}}{0.0034\,\text{atm/K}}$$
$$= 588\,K$$

$$588\,K - 273 = 315\,°C$$

So the valve would burst at a temperature between 310 °C and 320 °C. The external pressure is not the same as the internal pressure until the can bursts.

Wrap-up

Key Question: How does gas pressure change in flexible and in rigid containers?

- The relationship between gas pressure and temperature is expressed in the equation $P = kT$ and is called Gay-Lussac's law.
- When a gas sample is in a rigid container, its volume cannot change. When a gas sample is in a flexible container, its volume can change.
- In most containers, the temperature of the trapped gas changes to match the temperature of the gas on the outside.
- In flexible containers, the pressure of the gas on the inside changes to match the pressure of the air on the outside.

Evaluate (10 minutes)

> **Check-in**
>
> In the factory, a potato chip bag is filled with 50.0 mL of air. The pressure of the air is 1.0 atm, and the temperature is 25 °C. Imagine that you take the potato chips with you on an airplane. At higher altitudes, the air pressure in the cabin is 0.85 atm, and the temperature is 25 °C. The potato chip bag puffs up.
>
> 1. Which gas law applies?
> 2. Explain why the potato chip bag puffs up in the airplane.
> 3. What is the volume of the gas in the potato chip bag when it is at a higher altitude? Show your work.

Answers: 1. The temperature remains unchanged, but both the pressure and the volume change. You need to use the relationship between P and V (Boyle's law). 2. The bag puffs up because the air pressure outside the bag is lower than the air pressure inside the bag. 3. First, calculate k using P_1 and V_1. $k = P_1V_1 = 1.0$ atm • 50 mL = 50 mL • atm. Second, use k to determine V_2. $V_2 = k/P_2 = $ 50 mL • atm/0.85 atm = 59 mL.

Homework

Assign the reading and exercises for Weather Lesson 11 in the student text.

LESSON 12

OVERVIEW

Be the Molecule
Molecular View of Pressure

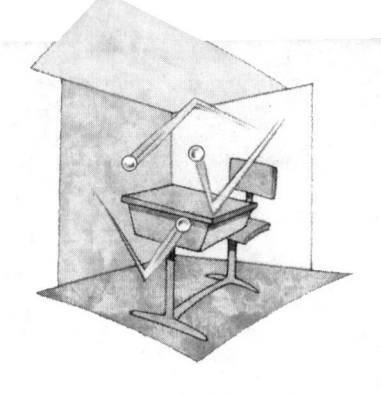

Lesson Type

Computer Activity:
  Whole Class
Demo:
  Whole Class

Key Ideas

Gas molecules are constantly in motion. They exert pressure on any object they collide with. Changing conditions, such as temperature and volume, can change the amount of pressure exerted by a gas. When gas molecules are squeezed into a smaller volume, they hit the walls of the container more often and thus exert more pressure. This relationship is expressed mathematically by the equation $k = PV$ (Boyle's law). When gas molecules are heated, they move faster and collide with the walls of their container more vigorously and more often. If the container is rigid, the volume cannot change, and the molecules exert more pressure. This relationship is expressed mathematically by the equation $k = P/T$ (Gay-Lussac's law). If the container is flexible, the volume changes so that the pressure remains the same. This relationship is expressed mathematically by the equation $k = V/T$ (Charles's law).

As a result of this lesson, students will be able to

- describe the motions of gas particles under changing conditions
- explain changes in pressure, volume, and temperature based on the motions of molecules

What Takes Place

In this lesson students examine a computer simulation that focuses on the effects of changing conditions on the pressure of a gas. Students discuss how to explain Charles's law, Boyle's law, and Gay-Lussac's law in terms of molecular speeds and collisions with the walls of the container. In an optional simulation, they have an opportunity to "be the molecule." This lesson reviews and reinforces the gas laws learned so far.

Materials

- student worksheet
- transparency—Be the Molecule (optional)
- computer and projector to display simulation
- 10 index cards or pieces of paper with the number 3, 4, or 5 written on them (for the simulation "Be the Molecule")

SciLinks

Topic: Chemistry Simulations
Visit: www.SciLinks.org
Web code: KEY-305

Demonstration materials

- 3 balloons
- sand for balloon
- water for balloon

Setup

Download a gas properties simulation from SciLinks that allows you to explore the gas laws by varying pressure, temperature, volume, and the number of particles.

Set up the projector to display the simulation and run through the simulation before class.

Prepare three balloons as a visual for the ChemCatalyst exercise. Fill one balloon with sand, one with water, and one with air.

Mark an area on the floor for the "Be the Molecule" simulation. You will need a space that is at least 10 ft. by 15 ft.

LESSON 12 GUIDE

Be the Molecule
Molecular View of Pressure

Engage (5 minutes)

Key Question: How do molecules cause gas pressure?

Show students balloons filled with sand, water, and air.

> **ChemCatalyst**
>
> Consider three balloons: one filled with sand, a second filled with water, and a third filled with air.
>
> 1. Describe at least three differences between the balloon containing air and the balloons containing sand and water.
> 2. What are the individual gas molecules doing inside the balloon containing air to make it big and round?

Sample Answers: 1. The balloons containing sand and water are much heavier. They are not round, and they do not maintain their shape. 2. The individual gas molecules must be hitting the inside of the balloon and pushing it out equally in all directions.

Discuss the ChemCatalyst

➟ Discuss differences among the three balloons.

Sample Questions

- Why don't the sand and water balloons maintain their shape?
- What makes it so hard to deform the air balloon?
- Why do gas molecules exert pressure in all directions?
- What is keeping the molecules in the liquid water from exerting pressure in all directions? (The motion of the molecules is restricted because the molecules are attracted to one another.)

Explore (15 minutes)

Introduce the Computer Simulation

➟ Tell students that they will view computer simulations of gas molecules and discuss their observations as a class.

106 *Living By Chemistry Teacher Guide*   Unit 3 Weather

Guide the Simulation

➡ Run Simulation 1: Constant Volume

1. Select constant volume for the simulation.
2. Select either the heavy species or light species and add gas particles to the container. (Selecting about 30 particles should be enough. You can experiment with more or less.)
3. Ask students to predict what will happen if you increase or decrease the temperature.
4. Use the heat control panel to increase or decrease the temperature.
5. Tell students to answer Question 1 on the worksheet.

➡ Run Simulation 2: Constant Temperature

1. Reset the simulation.
2. Select constant temperature.
3. Select either the heavy species or light species and add gas particles to the container. (Selecting about 30 particles should be enough. You can experiment with more or less.)
4. Ask students to predict what will happen if you increase or decrease the volume.
5. Click and drag on the person icon to increase or decrease the size of the container.
6. Tell students to answer Question 2 on the worksheet.

LESSON 12 ACTIVITY

Be the Molecule
Molecular View of Pressure

Name _____
Date _____ Period _____

Purpose
To examine how the motions of gas molecules cause gas pressure.

Part 1: Computer Simulations

1. For the first simulation, the volume does not change. Focus on what happens to the gas pressure as the temperature changes.

   a. What happens to the pressure when the temperature is increased? Explain why.

   Higher pressure. The molecules move at faster speeds. They hit the walls more often.
   (Note: The molecules also hit the walls with more force.)

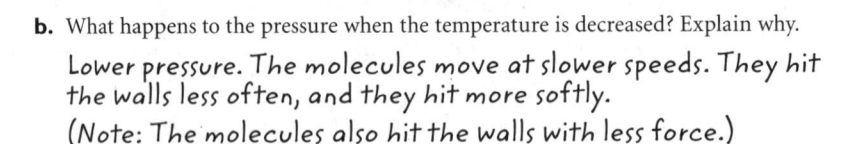

   b. What happens to the pressure when the temperature is decreased? Explain why.

   Lower pressure. The molecules move at slower speeds. They hit the walls less often, and they hit more softly.
   (Note: The molecules also hit the walls with less force.)

2. For the second simulation, the temperature does not change. Focus on what happens to the pressure as the volume of the container changes.

   a. What happens to the pressure when the volume is decreased? Explain why.

   Higher pressure. The molecules are squished into a smaller space. They move at the same speed, but they hit the walls more often because there is less distance between the walls.

   b. What happens to the pressure when the volume is increased? Explain why.

   Lower pressure. The molecules have a lot more space to move around in. They move at the same speed, but they hit the walls less often because there is a greater distance between the walls.

3. What conditions result in more collisions of molecules with the walls of the container and with one another?

   increasing the speed of the molecules; decreasing the size of the container.

4. Name two ways you could reduce the pressure of a gas sample.

   Lower the temperature. Increase the volume.

Part 2: Gas Law Review

1. **Fill in the table.** The first line of the table gives the volume, pressure, and temperature for a container of gas. The gas has an initial volume of 22.4 L. The pressure is 1.0 atm,

and the temperature is 300 K. Each subsequent row represents a new set of conditions for this gas. Fill in the blank spaces.

a.

Volume	Pressure	Temperature	Gas law
V_1 = 22.4 L	P_1 = 1.0 atm	T_1 = 300 K	(initial conditions)
11.2 L	1.0 atm	150 K	Charles's law
44.8 L	1.0 atm	600 K	Charles's law
89.6 L	1.0 atm	1200 K	Charles's law

b.

Volume	Pressure	Temperature	Gas law
V_1 = 22.4 L	P_1 = 1.0 atm	T_1 = 300 K	(initial conditions)
11.2 L	2.0 atm	300 K	Boyle's law
44.8 L	0.5 atm	300 K	Boyle's law
89.6 L	0.25 atm	300 K	Boyle's law

c.

Volume	Pressure	Temperature	Gas law
V_1 = 22.4 L	P_1 = 1.0 atm	T_1 = 300 K	(initial conditions)
22.4 L	0.5 atm	150 K	Gay-Lussac's law
22.4 L	2.0 atm	600 K	Gay-Lussac's law
22.4 L	4.0 atm	1200 K	Gay-Lussac's law

2. **Making Sense** In your own words, explain what gas pressure is and how it can be changed.

Gas pressure is the result of the gas molecules inside a container hitting the walls of that container. The more collisions, the greater the pressure. Also, the harder the collisions, the greater the pressure. There is more pressure when molecules are hot and moving really fast, or when a lot of molecules occupy a small space.

Explain and Elaborate (15 minutes)

Summarize the Gas Laws Studied So Far

⇒ You might want to display the computer simulations as you discuss Boyle's law, Charles's law, and Gay-Lussac's law.

⇒ Write the three gas laws on the board.

Sample Questions

- Describe a situation in which you would use Charles's law.
- Do any of the gas laws apply to a gas in an open container? Explain your thinking.
- What variable doesn't change for each gas law? (For Boyle's law, T doesn't change; for Charles's law, P doesn't change; for Gay-Lussac's law, V doesn't change.)
- What variable stays the same for all three gas laws? (amount of gas, or number of molecules)

Key Points

The gas laws allow you to calculate new values for gas temperature, pressure, and volume when *two* of these conditions change. With the gas laws discussed so far, it is understood that the gas is trapped or sealed off so that the number of molecules remains unchanged.

- **Charles's law:** $V = kT$, $k = V/T$ Pressure and amount of gas do not change.
- **Gay-Lussac's law:** $P = kT$, $k = P/T$ Volume and amount of gas do not change.
- **Boyle's law:** $P = k \cdot (1/V)$, $k = PV$ Temperature and amount of gas do not change.

Keep in mind that Charles's law and Gay-Lussac's law are proportional relationships: $V = kT$ and $P = kT$. Once you determine the proportionality constant, k, for one set of conditions for a specific gas sample, you can determine V and T, or P and T, for all other conditions for that gas provided the number of molecules doesn't change. Boyle's law is different, because P and V are inversely proportional: $P = k \cdot (1/V)$. But the mathematical manipulations are similar. Note that all temperatures must be converted to the Kelvin scale when you are solving gas law problems.

In the kinetic theory of gases, the gas molecules are in constant motion. They collide with the walls of the container. These collisions are the direct cause of the pressure on the walls of the container. When the temperature increases, the average speed of the molecules increases. When the molecules move faster, they collide with the walls more often and hit the walls harder. The result is higher gas pressure at higher temperature. Similarly, when the volume of a gas is reduced, the same number of molecules must occupy a smaller space. This means that they also collide with the walls of the container more often, resulting in higher pressure.

Student Simulation of Gas Molecules (optional)

[T] ⇒ Tell students you will be doing a simulation of the gas laws with the students acting as molecules. Assign ten students to be "molecules" and hand them each a card with the number 3, 4, or 5. Assign four students to watch the walls of the "container" and count the collisions. Display the transparency Be the Molecule and explain the rules of motion.

1. If you are playing the role of a gas molecule, you will have a speed card with a number from 3 to 5.
2. Find a tile on the floor anywhere inside the "container" as a starting point. Face one of the four "walls" of the container. This will be your initial direction of motion.
3. Each time your teacher says "go," move the number of tiles indicated by your speed card.
4. If you are about to collide with someone, exchange speed cards and move in the same direction the other person was moving before you "hit."
5. If you hit a wall, reverse your direction.
6. If you are assigned to count collisions, record each time your wall is hit by a person acting like a gas molecule.

⇒ Guide the student gas simulation for the three conditions listed. For each simulation, say "go" twenty times. Record the number of the wall collisions on the board at the end of each simulation.

1. Molecules move according to their speed cards.
2. All the speeds are doubled.
3. Molecules move according to their speed cards, but the volume of the container is cut in half.

	Collisions
1	
2	
3	

⇒ Ask students how this model could be made more realistic (gas molecules should move continuously, should start out moving in all different directions, etc.)

⇒ Ask students to summarize what they learned from the activity.

Wrap-up

Key Question: How do molecules cause gas pressure?

- On a molecular level, gas pressure is a result of gas molecules striking the walls of a container or an object.
- Pressure increases as the frequency of collisions increases. It also increases if the molecules hit the walls with greater force.
- Pressure and temperature are proportional if the volume of the gas does not change: $P = kT$. An increase in temperature increases both the collision frequency and the force with which the molecules hit the walls.
- Pressure and volume are inversely proportional if the temperature does not change: $P = k \cdot (1/V)$. A decrease in volume increases the frequency of collisions.

Evaluate (5 minutes)

> **Check-in**
>
> A family went for a drive in the desert. In the morning, the air pressure in the tires of their car was around 28 lb/in^2. In the afternoon, the tire pressure was around 32 lb/in^2. Provide an explanation on the molecular level for why this happened.

Answer: It is likely that the pressure increased because the daytime temperature increased and the tires got hot moving along the hot pavement. The gas molecules move faster as they heat up, increasing the pressure they exert. (*Note:* The volume of the tires may have increased a little, which would tend to decrease the pressure caused by the increase in temperature.)

Homework

Assign the reading and exercises for Weather Lesson 12 in the student text.

LESSON 13

What Goes Up
Combined Gas Law

OVERVIEW

Lesson Type
Classwork: Pairs

Key Ideas

A weather balloon rising into the atmosphere is affected by two variables, pressure and temperature, changing simultaneously. The temperature of the gas sample decreases as the balloon rises in altitude. The pressure of the gas also decreases as the balloon rises. The mathematical equation that describes the interactions of pressure, volume, and temperature together while the number of molecules remains the same is called the combined gas law. It is described mathematically by the equation $k = PV/T$. This equation can be used to predict the volume of the weather balloon under various conditions.

As a result of this lesson, students will be able to

- define the combined gas law
- solve gas law problems that involve changes in all three of the variables, P, V, and T

Focus on Understanding

- The math that accompanies this gas law is a bit trickier than the math associated with the three gas laws discussed previously. We've taken an approach that focuses on the proportionality constant, k.
- Students might have difficulty grasping the concept that the temperature and pressure inside the balloon will match the temperature and pressure outside the balloon.

Key Term

combined gas law

SciLinks NSTA
Topic: Gas Laws
Visit: www.SciLinks.org
Web code: KEY-306

What Takes Place

Students are introduced to weather balloons and the combined gas law. They complete a worksheet focusing on the ascent of a weather balloon through the changing atmosphere. They calculate how the volume of the gas sample changes with increasing altitude, using the equation $k = PV/T$. They wrestle with the question of which has a greater effect on the volume of the balloon, the decreasing air temperature or the decreasing air pressure.

Materials

- student worksheet
- transparency—Altitude Table

LESSON 13 GUIDE

What Goes Up
Combined Gas Law

Engage (5 minutes)

Key Question: What is the relationship among pressure, volume, and temperature for a sample of gas?

> ### ChemCatalyst
> A weather balloon is inflated with helium to a volume of 125,000 L. When it is released, it rises high into the atmosphere, where both the pressure and the temperature are lower.
>
> 1. Explain why the balloon rises.
> 2. Will the balloon pop at a high altitude? Explain your thinking.

Sample Answers: 1. The gas in the helium balloon is less dense than the surrounding air, because the mass of the balloon (He) is less per unit volume than the mass of the air molecules (mostly O_2 and N_2). 2. When a sample of gas decreases in temperature, it contracts (Charles's law). When a sample of gas decreases in pressure, it expands (Boyle's law). Because these are opposing effects, it is difficult to predict what will happen to the balloon from what we know at this point. It depends on whether temperature changes or pressure changes are more important.

Discuss the ChemCatalyst

➡ Assist students in sharing their ideas about how temperature and pressure changes affect a weather balloon.

Sample Questions

- How do pressure and temperature affect the volume of the balloon?
- How would you explain the observation that a weather balloon has burst at some altitude and fallen back to Earth?
- How do you think a weather balloon can be used to probe the atmosphere?

Explore (20 minutes)

Introduce the Classwork

➡ Tell students that they will be working with pressure, temperature, and volume data for a weather balloon. Briefly introduce weather balloons and atmospheric changes with altitude to the class.

➡ Tell students that weather balloons are large balloons filled with helium or hydrogen gas that carry instruments for studying the atmosphere at high altitudes. As the balloon rises, the instruments take measurements. This information is relayed back to a station via a transmitter. Unlike a hot air balloon, a weather balloon is a closed container.

- Remind students that the atmosphere changes in pressure and temperature with altitude. In general, both these variables decrease with altitude. A gas sample in the air will respond to these changes.

- Introduce the combined gas law, $k = \frac{PV}{T}$, and complete a sample problem. Then have students work in pairs.

Example

A balloon has a volume of 12,000 L at sea level where the pressure is 1.0 atm and the temperature is 285 K. The balloon is released and travels to an altitude of 5,000 feet where the temperature is 278 K and the pressure is 0.80 atm. What is the new volume of the balloon?

Solution

$$k = \frac{P_1 V_1}{T_1}$$

$$= \frac{1.0 \text{ atm} \cdot 12{,}000 \text{ L}}{285 \text{ K}}$$

$$= 42.1 \frac{\text{atm} \cdot \text{L}}{\text{K}}$$

$$k = \frac{P_2 V_2}{T_2}$$

$$42.1 \frac{\text{atm} \cdot \text{L}}{\text{K}} = \frac{(0.80 \text{ atm})(V_2)}{278 \text{ K}}$$

$$V_2 \approx 15{,}000 \text{ L}$$

LESSON 13 CLASSWORK
What Goes Up
Combined Gas Law

Name _____
Date _____ Period _____

Purpose

To explore what happens when the temperature, pressure, and volume of a gas all change at the same time.

Tracking the Volume of a Weather Balloon

A weather balloon is filled to a volume of 12,500 L at sea level. The air pressure is 1.0 atm and the temperature is 290 K. These starting values are listed in the bottom row of the table.

Altitude (ft)	Temperature (°C)	Temperature (K)	Pressure (atm)	Volume (L)
40,000 ft	−57 °C	216 K	0.20 atm	47,000 L
30,000 ft	−45 °C	228 K	0.30 atm	33,000 L
25,000 ft	−35 °C	238 K	0.40 atm	26,000 L
10,000 ft	−5 °C	268 K	0.70 atm	17,000 L
5,000 ft	5 °C	278 K	0.80 atm	15,000 L
0 ft	17 °C	290 K	1.0 atm	12,500 L

1. What is the value of *PV/T* at these beginning conditions?

   The value of PV/T is 43.1 atm · L/K.

2. **Example:** The balloon is released and travels to an altitude of 5,000 ft. Here is how the value of *PV/T* can be used to calculate the volume of the balloon under these new conditions:

$$k = \frac{P_1 V_1}{T_1} = \frac{(1.0 \text{ atm})(12,500 \text{ L})}{290 \text{ K}} = 43 \frac{\text{atm} \cdot \text{L}}{\text{K}}$$

$$k = \frac{P_2 V_2}{T_2}$$

$$43 \frac{\text{atm} \cdot \text{L}}{\text{K}} = \frac{(0.80 \text{ atm})(V_2)}{278 \text{ K}}$$

$$V_2 = 15,000 \text{ L}$$

   Write this new volume in the table.

3. Did the volume of the balloon increase or decrease when it rose to 5,000 ft? Explain why.

   The volume increased due to the decrease in pressure. So the decrease in pressure had a larger effect on the volume than did the decrease in temperature.

4. Calculate the rest of the values that are missing in the table. Fill in the table with your answers.

5. What is the value of *PV/T* at sea level? at 25,000 ft? Explain why this number is useful in your calculations.

   The value of PV/T is always 43 for this balloon. This is the proportionality constant, and it relates P, V, and T to one another mathematically.

6. Suppose a second weather balloon is filled with 25,000 L of helium at 290 K and 1.0 atm. What is the value of the proportionality constant for this balloon?

   The value of k, the proportionality constant, would be different. It would be (1.0 atm)(25,000 L)/290 K, or 86 atm · L/K.

7. **Making Sense** What happened to the volume of the balloon as it rose? What explanation can you offer for this?

   The volume of the balloon increased as it rose because the pressure changes caused the balloon to expand more than the temperature changes caused the balloon to contract.

8. **If You Finish Early** Suppose a weather balloon is designed to burst when the volume expands to 40,000 L.

   a. For a balloon filled at sea level with 12,500 L of helium, use the table to estimate the altitude at which the balloon will burst.

   According to the table, the volume for this balloon is 33,000 L at 30,000 ft and 47,000 L at 40,000 ft. Because 40,000 L is between these two numbers, the balloon will burst at about 35,000 ft.

   b. If this same balloon is filled at sea level with 25,000 L of helium, will it burst at the same altitude, a higher altitude, or a lower altitude than if it started with 12,500 L of helium? Support your answer with a calculation.

   The balloon starts off bigger, so it will reach 40,000 L at a lower altitude, probably closer to 20,000 ft. At 20,000 ft, the pressure is roughly 0.5 atm and the temperature is roughly 250 K. Here is the calculation:

   $$\frac{PV}{T} = k, \text{ the proportionality constant}$$

   Use the original values of P, T, and V to solve for k:

   $$k = \frac{(1.0 \text{ atm})(25,000 \text{ L})}{290 \text{ K}} = 86 \frac{\text{atm} \cdot \text{L}}{\text{K}}$$

   Solve for the new volume, V:

   $$k = 86 \frac{\text{atm} \cdot \text{L}}{\text{K}} = \frac{(0.5 \text{ atm})(V)}{250 \text{ K}}$$

   $$V = 43,000 \text{ L}$$

Explain and Elaborate (15 minutes)

Introduce the Combined Gas Law

→ Write the combined gas law on the board.

> **Combined Gas Law**
> $$k = \frac{PV}{T} \quad \text{or} \quad \frac{P_1 V_1}{T_1} = \frac{P_2 V_2}{T_2}$$

Sample Questions

- If the temperature and volume of a gas change, can you determine the pressure?
- If the volume of a gas increases and the pressure decreases, can you determine the temperature?
- If you change the temperature of the gas in a balloon, will the value of *PV/T* remain the same? Explain.
- Suppose you have two balloons filled with different amounts of gas. Will the value of *PV/T* be the same? Explain.

Key Point

The relationship among the pressure, temperature, and volume of a gas is described by the combined gas law. It states that the value of *PV/T* stays the same for a set of conditions for a given sample of gas. This law allows you to calculate one variable when the other two change.

Explore How Changing *T* and *P* Affects the Weather Balloon

[T] → Use the transparency Altitude Table to assist in the discussion.

Sample Questions

- Does a decrease in air temperature have an effect on the gas pressure in the balloon? Explain.
- The weather balloon continues to rise no matter what conditions we impose on it. Why? (It's full of helium.)
- Why do you think the balloon eventually pops?
- Which change has a greater effect on the volume of the balloon when you increase in altitude, the air temperature or the air pressure?

Key Point

The weather balloon rises no matter what the outside conditions because the balloon is full of helium, and helium is less dense than air. With a decrease in temperature, we would expect the balloon to decrease in volume (Charles's law). However, with a decrease in pressure, we would expect the balloon to increase in volume (Boyle's law). It appears that the changing air pressure has a greater effect on the volume of the balloon than does the changing air temperature. The balloon increases in volume until the pressure inside is too great for the material the balloon is made of and the balloon pops.

Wrap-up

Key Question: What is the relationship among pressure, volume, and temperature for a sample of gas?

- If volume, temperature, and pressure all vary, then you can use the combined gas law to determine the effects of changing two variables on the third. (Amount of gas remains the same.)

$$k = \frac{PV}{T}$$

- Air temperature and air pressure both decrease with increases in altitude.
- Ultimately, for a weather balloon, gas pressure has a greater effect on gas volume than does gas temperature.

Evaluate (5 minutes)

Check-in

A sample of neon gas occupies a volume of 1.0 L at 300 K and 1.0 atm.

1. Calculate the value of the proportionality constant, k.
2. Suppose you increase the temperature to 600 K and decrease the pressure to 0.50 atm. Does the volume of the gas increase or decrease? Explain your answer.

Answers

1. $k = \frac{P_1 V_1}{T_1} = \frac{1.0 \text{ L} \cdot 1.0 \text{ atm}}{300 \text{ K}} = 0.0033 \ \frac{\text{atm} \cdot \text{L}}{\text{K}}$

2. The volume definitely will be larger under the new conditions.

$$k = \frac{P_2 V_2}{T_2}$$

$$0.0033 \ \frac{\text{atm} \cdot \text{L}}{\text{K}} = \frac{0.50 \text{ atm} \cdot V}{600 \text{ K}}$$

$$= 4.0 \text{ L}$$

Homework

Assign the reading and exercises for Weather Lesson 13 in the student text.

LESSON 14 OVERVIEW

Cloud in a Bottle
High and Low Air Pressure

Key Ideas

Meteorologists regularly track the pressure of the air around us in order to better forecast the weather. From air pressure measurements, high- and low-pressure areas are charted on weather maps. Low-pressure areas are associated with clouds and storms, while high-pressure areas are associated with sun and clear skies.

As a result of this lesson, students will be able to

- explain the influence of high- and low-pressure systems on the weather

Lesson Type
Demo:
  Whole Class
Lab:
  Groups of 4

What Takes Place

This lesson applies what has been learned about gas behavior to the weather context. Students watch a demo of volume change involving temperature and phase changes. Then, working in groups, students make a "cloud in a bottle." During the discussion, students make connections between high and low pressure and weather.

Materials

- student worksheet
- vacuum chamber and vacuum pump (optional/recommended)

Demonstration materials

- 250 mL Erlenmeyer flasks (2)
- 2 medium-size party balloons
- hot plate
- oven mitt
- ice and water

Per group of 4

- 2 L empty, clear plastic soda bottle with cap
- warm tap water
- long safety matches

Setup

Practice the demo and making a cloud in a bottle. For making a cloud, a vacuum chamber and vacuum pump works best, but an empty plastic soda bottle also works.

LESSON 14 GUIDE

Cloud in a Bottle
High and Low Air Pressure

Engage (5 minutes)

Key Question: How are areas of high and low air pressure related to the weather?

ChemCatalyst

Clouds are tiny water droplets suspended in the air.

1. Are pressure, temperature, or volume changes involved in the formation of clouds? Explain your thinking.
2. Cloud formation is related to low pressure. Explain why.

Sample Answers: 1. Students might speculate that clouds form when moisture in the air gets cold enough to undergo a phase change. 2. If water condenses, there are fewer gas molecules in the air. Perhaps fewer gas molecules means lower pressure.

Discuss the ChemCatalyst

Sample Questions

- How does water vapor get into the atmosphere? Why does water vapor rise?
- What causes water vapor to condense to form clouds?

Explore (20 minutes)

Complete the Demonstration

➡ Demonstrate how evaporation is associated with high pressure and condensation is associated with low pressure.

1. Draw students' attention to a 250 mL Erlenmeyer flask with 5 mL of water in it. Place a balloon over the mouth of the flask.
2. Heat the flask on a hot plate. Do not let all the water boil away.
3. After several minutes, use an oven mitt to remove the flask from the hot plate.
4. Solicit students' observations.
5. Ask why the balloon gets so large. The number of gas molecules increases as water evaporates.

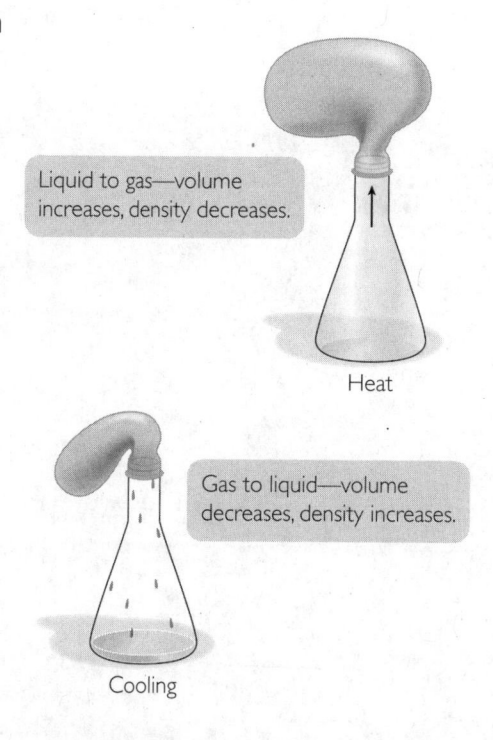

Liquid to gas—volume increases, density decreases.

Heat

Gas to liquid—volume decreases, density increases.

Cooling

Section II   Lesson 14 Cloud in a Bottle   **121**

→ Repeat the procedure, but this time attach the balloon after removing the flask from the hot plate.

1. Place the flask upright in ice water.

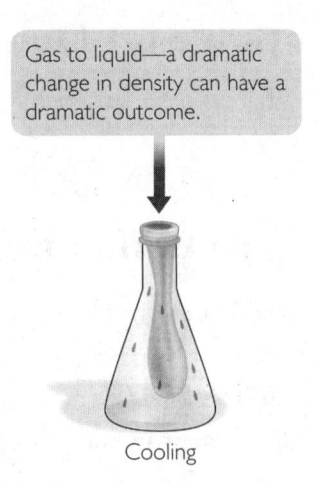

> Gas to liquid—a dramatic change in density can have a dramatic outcome.

Cooling

2. Solicit students' observations.

3. Ask why the balloon goes inside the flask. (Before the balloon was placed on the flask, the air in the flask was mainly water vapor. When this water vapor condenses, the pressure decreases inside the flask. The air on the outside pushes the balloon into the flask.)

Introduce the Lab

→ Tell students they will be using the pressure-temperature relationships to create a cloud inside a bottle. (If the cloud in the bottle is too difficult to see and you have the apparatus, you can demonstrate the cloud in the bottle with a vacuum pump and vacuum chamber instead. Note that this will change the situation as V will remain the same while n changes. Changes in n are covered in the next section.)

→ Students can work in groups of 4.

LESSON 14 LAB

Cloud in a Bottle
High and Low Air Pressure

Name _____
Date _____ Period _____

Purpose
To find the connection between air pressure and the weather forecast.

Part 1: Cloud in a Bottle
Materials
- 2 Liter plastic soda bottle with cap
- warm tap water
- long safety matches

Safety Instructions
⚠ Safety goggles should be worn at all times.

Procedure
1. Put a small amount (about 50 mL) of warm water into the plastic bottle.
2. Light a match. Blow it out and then hold it inside the bottle to collect some smoke.
3. Quickly remove the match and put the cap tightly on the bottle.
4. Shake the bottle to add moisture to the air inside.
5. Squeeze the bottle firmly, then release. Repeat. Observe the air inside the bottle.
6. Repeat the experiment, this time with 50 mL of cold water.
7. Next, repeat the experiment with a dry bottle. Do not add water. Simply create smoke, close the bottle, and squeeze and release.

Observations
1. What did you observe inside the bottle when you squeezed and released the bottle?

    A faint cloud forms inside the bottle

2. What gas law was operating during this experiment? Explain.

    Sample Answer: Gay-Lussac's law, because pressure and temperature are changing but volume stays nearly the same. (Or: Combined gas law, because P, V, and T are changing but n remains the same.)

3. If P decreases and V increases when the bottle is released, what do you think happens to T? What evidence do you have?

    The temperature must decrease because the water vapor in the air undergoes a phase changce, turning to liquid water droplets.

4. What happened when you used cold water in the bottle?

    Sample Answer: There was less of a cloud.

Living By Chemistry Teaching and Classroom Masters: Units 1–3
© 2010 Key Curriculum Press

Unit 3 Weather 241
Lesson 14 • Worksheet

5. What did you observe when you used a dry bottle?

   No real cloud forms. There is not enough moisture in the air for a cloud to form.

6. Low-pressure areas are the result of air rising into the atmosphere from Earth's surface. Explain how this might result in cloud formation over a low-pressure area.

   Sample Answer: Moist air moving up is cooled at higher altitudes. This air condenses to form clouds.

7. High-pressure areas are the result of air falling from high altitudes and expanding. Explain how this might result in clear skies over a high-pressure area.

   Sample Answer: Moisture is not moving upward, so clouds cannot form.

8. **Making Sense** What did you learn about cloud formation from today's activity?

   Clouds form when water vapor changes phases. It helps to have particles in the air, like smoke. Changing pressure and temperature cause gaseous water to change phase.

9. **If You Finish Early** Meteorologists sometimes measure the air pressure in millibars. Suppose the air pressure is 980 millibars for a given location. Convert the air pressure to atmospheres of pressure. (1 atm = 1013 mb.)

   980 millibars is 0.97 atm

Explain and Elaborate (15 minutes)

Process the Lab

Sample Questions

- What variables did you change when you pumped the air out of the bottle?
- What caused the cloud to form?
- Do you think the water vapor in the air in the room had any effect on the cloud formation?

Key Points

In the plastic bottle, pressure and temperature were changing, but the volume stayed (nearly) the same. This situation is covered by Gay-Lussac's law, $P = kT$. When the air pressure inside the bottle is lowered, the temperature inside the bottle also decreases. This decrease in temperature causes water vapor in the bottle to change phase and form droplets. The smoke provides some particles in the air for the water droplets to cling to or form around, which can be seen as a faint cloud inside the bottle.

Gay-Lussac's law also tells us that when the pressure inside the bottle is increased, the temperature also must increase. This is why the cloud evaporates again when you squeeze the bottle.

Discuss Cloud Formation in Earth's Atmosphere

Sample Questions

- Is the air in our atmosphere in a container? Explain.
- What are the first steps in the formation of clouds? (evaporation of water, followed by decreases in pressure and temperature)
- Why does water vapor rise?
- What happens to the temperature and pressure of the water vapor as it rises?

Key Points

Clouds form when water vapor in the atmosphere changes phase and forms water droplets. Water vapor enters the atmosphere due to evaporation from bodies of water. When the water is warmer, more water evaporates. This is one reason why warm, tropical areas have lots of rainfall. Because no "container" is associated with Earth's atmosphere, it is easier to explain weather by focusing on pressure and temperature changes of air masses.

Warm air rises because it is less dense than cold air. As warm air rises, the temperature and pressure decrease. This causes the water vapor in the air to condense into droplets, forming clouds. A low-pressure system is created near Earth's surface when warm air rises up into the atmosphere. A high-pressure system is created near Earth's surface when cool air sinks from high in the atmosphere.

Enough water vapor must be present in the air in order for clouds to form. This is why more clouds form over the rainforest than over the desert. Fog is common near bodies of water. On warm days, when the air has the potential to hold more moisture, large puffy clouds can be observed at lower altitudes. On cold days, when there is less moisture in the air, the clouds tend to be wispy and high in the sky.

Wrap-up

Key Question: How are areas of high and low air pressure related to the weather?

- Clouds form when both air pressure and temperature are low.
- Areas of low air pressure are associated with fronts, storms, clouds, and precipitation, whereas areas of high air pressure are associated with clear skies and pleasant conditions.

Evaluate (5 minutes)

> **Check-in**
>
> On a camping trip you take a sealed plastic water bottle to a higher elevation. When you arrive on the mountain you notice the bottle is slightly larger and condensation has formed on the inside of the bottle. Explain what happened in terms of P, V, T, and the quantity PV.

Answer: The volume has increased slightly because the atmospheric pressure is lower. So the pressure inside the bottle has increased. Since water condensed, the temperature must have decreased. The quantity of PV must have decreased since the ratio PV/T is a constant.

Homework

Assign the reading and exercises for Weather Lesson 14 and the Section II Summary in the student text.

III Concentrating Matter

The third section of the Weather unit focuses on the number of gas particles in a sample and how this is related to pressure, temperature, and volume. Lesson 15 explores the atmosphere and how the density of gas molecules varies with altitude. In Lesson 16, students learn about standard temperature and pressure and how the number of particles is related to the volume of a gas. Lesson 17 is a lab in which students measure the volume of an average breath and then calculate the number of gas particles in that volume using the ideal gas law. Lesson 18 extends the concept of number density to water vapor as students learn about relative humidity. Extreme weather conditions are the subject of Lesson 19. Lesson 20 is a review of the entire Weather unit.

In this section, students will learn

- about pressure and temperature variations in the atmosphere
- the relationship between the pressure and the number density of a gas
- about the unit the mole
- Avogadro's hypothesis and the ideal gas law
- about water vapor density and humidity
- about the formation of hurricanes

LESSON 15 OVERVIEW

n Is for Number

Pressure and Number Density

Lesson Type
Lab: Groups of 4

Key Ideas

The density of molecules in the atmosphere can be expressed as n/V, where n is the number of gas particles and V is the unit volume. This is known as the number density (as opposed to the "mass density"). As n/V increases, the pressure of the gas also increases, and vice versa. When the temperature remains the same, the pressure of a gas is directly proportional to the number density of the gas: $P = k \cdot (n/V)$. In our atmosphere, the value of n/V decreases as you go up in altitude, and therefore the pressure, P, decreases.

As a result of this lesson, students will be able to

- define the number density of a gas
- describe the number density of the atmosphere as it relates to altitude
- explain the relationship between number density and gas pressure
- describe one way to measure gas pressure

Focus on Understanding

- Students might wonder why this particular mathematical relationship does not have a formal gas law name. You could provide them with an easy way to refer to it, such as the number density equation or the number/pressure relationship.

Key Terms

number density
barometer

What Takes Place

This lesson examines the effect of the number of air molecules or atoms on air pressure. Students are provided with a handout that summarizes the changes in Earth's atmosphere with altitude. Part 1 of the worksheet focuses on how changing the number of particles, n, affects gas pressure in our atmosphere. Part 2 contains an activity in which students are challenged to change the water level in a U-tube by manipulating pressure, volume, or the number of gas particles. This leads to the idea that the difference in height of the liquid in the U-tube reflects the atmospheric pressure. The discussion focuses on how number density, n/V, is related to gas pressure.

Materials

- student worksheet
- transparency—ChemCatalyst
- handout—Earth's Atmosphere

Per group of 4
- approximately 3 ft of clear tubing (about 5/16 in. inner diameter to fit on syringe)
- 2 ring stands
- 2 burette clamps
- wash bottle
- uncapped large syringe
- balloon
- cork
- plastic tub or sink
- water

Setup

Students will put about 30 mL of water in the tubing. There may be an overflow of water as students manipulate the U-tubes. Set up catch basins or have students work over a sink. Note that it is very easy to make water squirt some distance out of the U-tube using the syringe. Plan accordingly for this possibility. Have one setup available for the Explain and Elaborate discussion following the lab.

LESSON 15 GUIDE

n Is for Number
Pressure and Number Density

Engage (5 minutes)

Key Question: How is the number of gas molecules in a sample related to pressure?

ChemCatalyst
Earth's Atmosphere

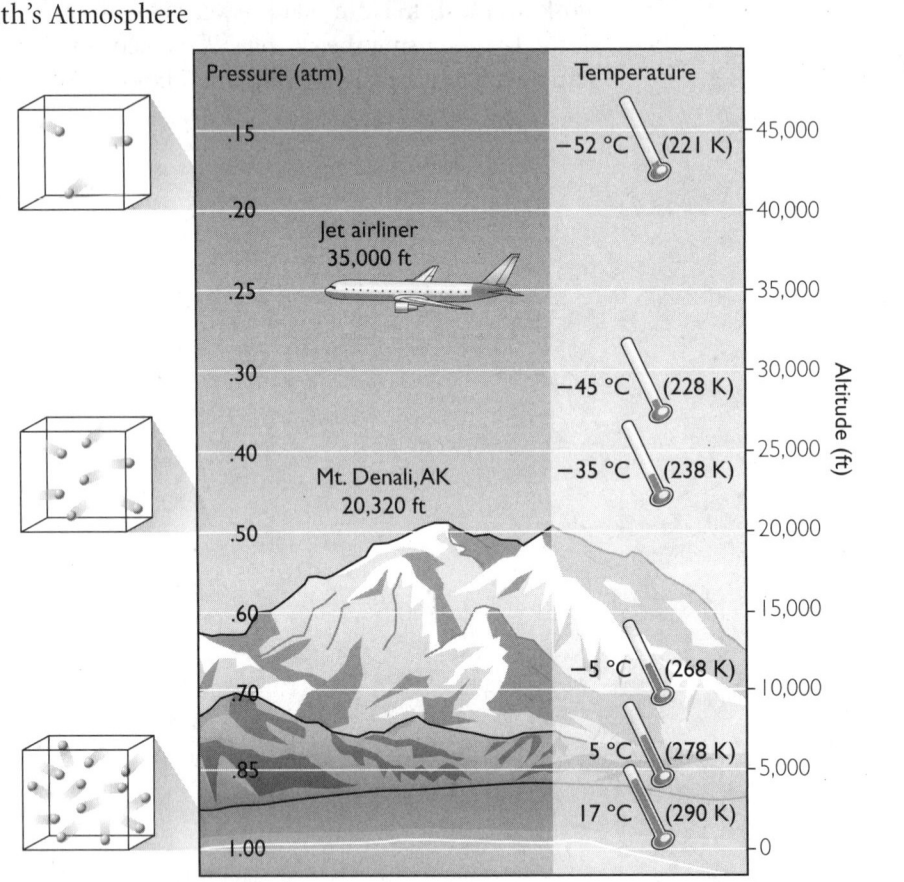

Compare the atmosphere at sea level and at 34,000 ft, the altitude at which airplanes fly.

1. Describe at least three differences.
2. Explain why it is difficult to breathe at 34,000 ft.

Sample Answers: 1. The temperature and air pressure are lower, and there are fewer molecules per unit of volume, at 34,000 ft. 2. Each breath you take at 34,000 ft contains only one-quarter as many air molecules as air at sea level.

Discuss the ChemCatalyst

⇒ Discuss air pressure in the atmosphere.

Sample Questions

- Explain how air changes as altitude increases.
- How does the amount of oxygen in one breath at 34,000 ft compare to the amount in one breath at sea level? Explain your thinking.
- Use the kinetic view of gases to explain how the number of gas molecules is related to pressure.

Explore (15 minutes)

Introduce the Lab

⇒ Pass out the handout Earth's Atmosphere.
⇒ Tell students that they will answer a set of questions for Part 1. Then briefly explain the procedure for Part 2. Ask students to work in groups of four.
⇒ Tell students to wear safety goggles.

Hold up the U-tube equipment for students to see. Tell students they will be challenged to change the water level inside the tube using only the materials provided.

Remind students that the air pressure of the atmosphere is pressing down on the water in both sides of the open tube.

Tell students that the letter *n* refers to the number of atoms or molecules being considered in a sample of gas.

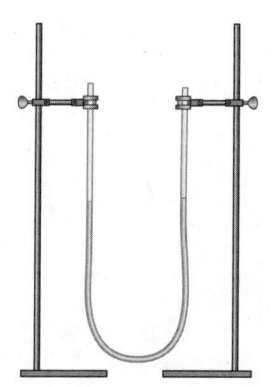

Section III
Lesson 15 *n* Is for Number   131

LESSON 15 LAB

n Is for Number
Pressure and Number Density

Name _____
Date _____ Period _____

Purpose
To explore how number density affects gas pressure.

Materials
- handout—Earth's Atmosphere
- 2 ring stands
- 2 burette clamps
- approximately 3 ft of clear tubing
- wash bottle
- uncapped large syringe
- balloon
- cork
- water

Safety Instructions
⚠ Wear safety goggles for Part 2.

Part 1: The Atmosphere
Use the handout on Earth's atmosphere to help you answer these questions.

1. Give two reasons why the air pressure decreases as the altitude increases.

   As altitude increases there are fewer molecules and the temperature is lower, so the molecules are moving slower.

2. Use the kinetic theory of gases to explain why the gas pressure increases as the number of gas molecules in a container increases. Assume that the temperature does not change.

   Pressure is caused by collisions of the molecules with the walls of the container. If there are more molecules in the same volume moving at the same average speed, the pressure increases because more collisions with the walls of the container occur.

3. The illustrations show three samples of air at the same temperature. List the samples in order of increasing gas pressure. Explain your reasoning using the kinetic theory of gases.

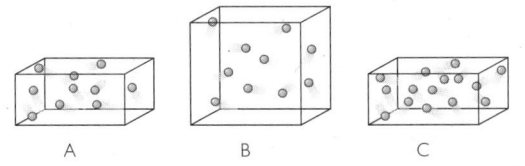

   A   B   C

   The correct order from low to high pressure is B, A, C.
   Container B has the least number of molecules compared to its volume. If we compare equal volumes from all three containers, Container C has the largest number of molecules per unit volume.

4. The number of gas molecules per unit of volume (such as 1 cm^3) is called the number density of a gas. List the three samples of air in Question 3 in order of increasing number density. *The correct order is B, A, C.*

Living By Chemistry Teaching and Classroom Masters: Units 1–3
© 2010 Key Curriculum Press

Unit 3 Weather 243
Lesson 15 • Worksheet

132 *Living By Chemistry Teacher Guide* Unit 3 Weather

Part 2: Balancing Air Pressure

Procedure and Questions

1. Hold the two ends of a piece of plastic tubing so that it forms a U shape with both ends pointing up.
2. Use a wash bottle to fill the tubing with water so that it reaches a level about halfway up each side.
3. Use two ring stands with burette clamps to hold the tubing when you want to free up your hands.
4. Find three ways to get the water levels to stay at different heights. Do not use your mouth on the tubing. You can use the syringe, cork, and balloon if you wish. Describe or draw a diagram of each.

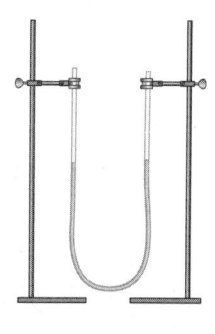

   Sample answers: Seal off one end with a syringe, leave the other end open, and then depress the plunger. Seal off one end with a finger and raise the sealed end above the open end. Seal off one end with a finger and lower the sealed end below the open end.

5. In general, is there greater air pressure on the side where the water is higher or on the side where it is lower? Explain your thinking.

   There is greater air pressure on the side where the water is lower. The air is pushing down the water on this side, causing the water on the other side to rise.

6. Is the number density of the gas, n/V, greater on the side where the water level is higher or on the side where it is lower? Explain your thinking in terms of the kinetic theory of gases.

   On the side where the water level is lower, the pressure is higher, so the number density, n/V, is higher. More molecules are colliding with the walls.

7. Suppose you have water levels at equal heights. You put a stopper on one side and leave the other side open. The next day, the water level on the open side is lower. Has a high or low air pressure system moved in? Explain.

   A high-pressure system has moved in. The air pressure in the atmosphere is higher and is pushing down on the water.

8. **Making Sense** Explain how air pressure is related to the number density of a gas.

   As the number density of a gas increases, there are more gas molecules in a given volume. The pressure increases.

EARTH'S ATMOSPHERE

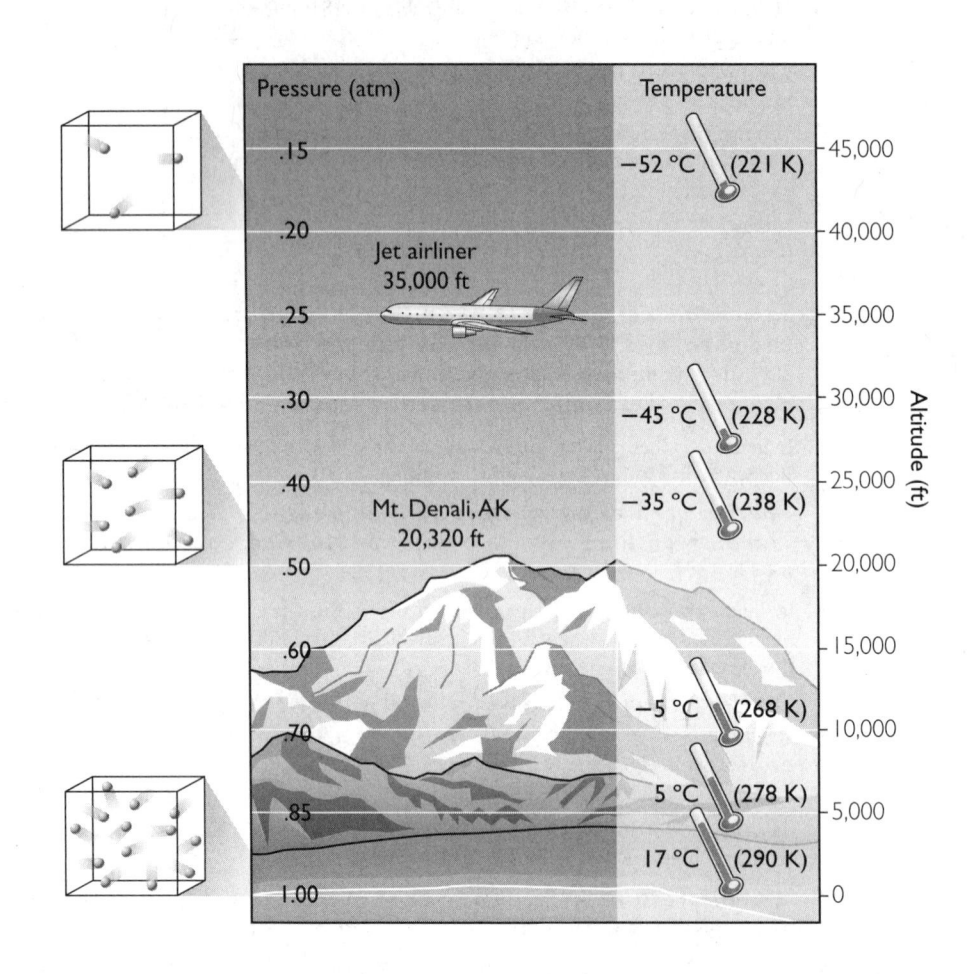

Composition of dry air:

Main gases:
$N_2(g)$: 78.084%
$O_2(g)$: 20.946%
$Ar(g)$: 0.934%

Other gases:
$CO_2(g)$: 0.035%
$Ne(g)$: 0.0018%
$He(g)$: 0.0005%

$CH_4(g)$: 0.00017%
$Kr(g)$: 0.0001%
$H_2(g)$: 0.000055%

Explain and Elaborate (15–20 minutes)

Discuss the Relationship Between Pressure and Number Density

➡ Write the following on the board: The number of gas molecules per unit of volume, n/V, is called the number density.

➡ At the appropriate point in the discussion, write the proportional relationship $P = k \cdot (n/V)$ on the board.

Sample Questions

- Use the kinetic theory of gases to explain why P increases as n increases.
- Use the kinetic theory of gases to explain why P increases as n/V increases.
- Why is it useful to talk in terms of changing number density, n/V, rather than just changing number, n?

Key Point

The gas pressure increases as the number of gas molecules per unit of volume increases. If you add more gas molecules to a container that doesn't change size, the pressure of the gas increases because there are more molecules colliding with the walls of the container. The relationship between P and n/V is proportional.

$$P = k\left(\frac{n}{V}\right)$$

Note that it usually is more useful to consider how n/V changes rather than simply n, especially in a flexible container with variable volume. Pumping up a bicycle tire is an example of this. The bicycle pump puts more and more air molecules into the tire, increasing the number density and the air pressure inside the tire.

> **Number density:** The number of gas particles per unit volume.
> Number density $= \frac{n}{V}$

Discuss Students' Solutions to the Challenge

➡ Have a U-tube with water available at the front of the room for demonstration.

➡ Ask student groups to share, draw a picture of, or demonstrate how they managed to change the water levels in the U-tube.

Sample Questions

- What happens to the water levels when the U-tube is open to the atmosphere on both ends?
- Explain how your group managed to make the water levels inside the tube uneven.
- Explain what you did in terms of changing pressure and number density.

Key Points

The height of the water levels in a U-tube indicates differences in pressure. When the U-tube is open to the atmosphere at both ends, the water levels are always at the same height, even if one end of the tubing is higher than the other. This is because the gas pressure from the atmosphere is the same on both sides.

In order to make the water levels uneven, it is necessary to seal off one end with a finger, a syringe, a cork, or a balloon. This allows you to manipulate the volume of the gas molecules on the side that is sealed off and thereby change the number density, n/V, relative to the number density of the atmosphere. The side with the lower water level has the higher pressure.

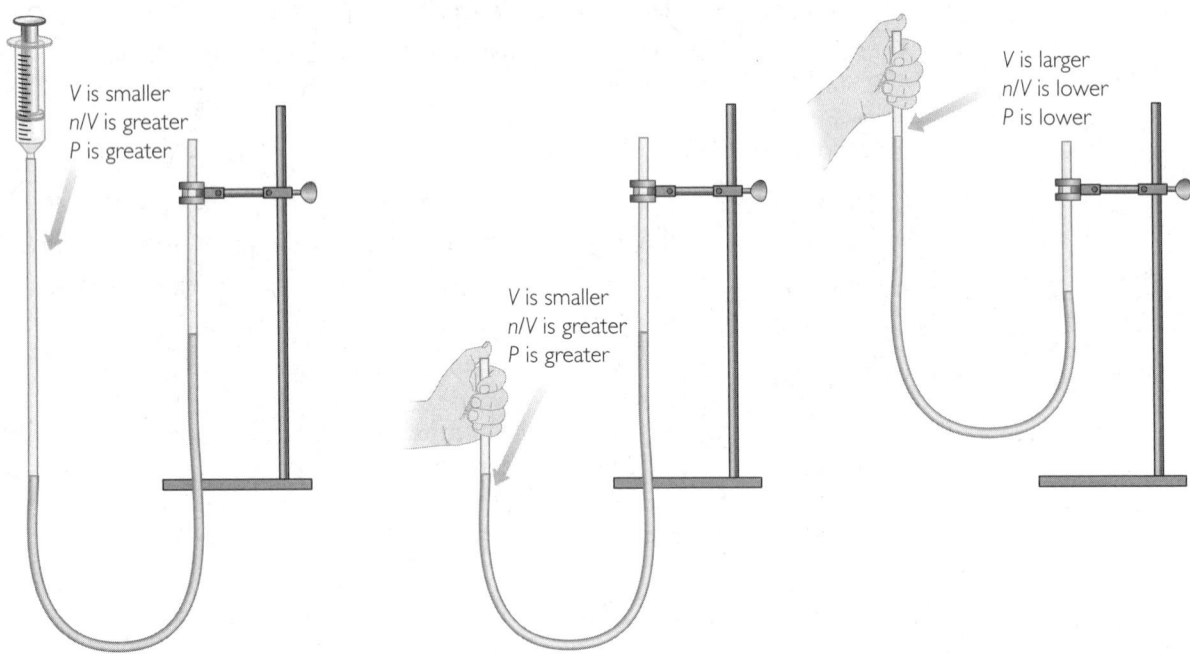

V is smaller
n/V is greater
P is greater

V is smaller
n/V is greater
P is greater

V is larger
n/V is lower
P is lower

Consider the syringe as part of the system. Depressing the plunger reduces the volume while n stays the same. This increases n/V, which increases the pressure P on that side. You can tell that the pressure is greater on the side with the syringe, because the trapped air is pushing up the water on the other side.

Sealing off one end means n will stay the same on that side. When you lower that end, the weight of the water compresses the air on that side, causing the gas volume to decrease; n/V increases, so P increases.

If you seal off one end and then raise that end, the number of molecules, n, stays the same on the side that is sealed off. The weight of the water makes it move down in the tubing, resulting in a slightly larger gas volume. This decreases n/V, and therefore P, on the sealed-off side.

Discuss Measurement of Atmospheric Pressure

Sample Questions

- Suppose you have a U-tube partially filled with water. One end has a stopper on it, and the other end is open. What happens to the water levels if you take the U-tube up a mountain?
- What happens to the water levels in this same U-tube if a high-pressure system moves into a region?

Key Point

Air pressure can be determined by measuring the difference in height of a liquid. The U-tube can be used to measure the air pressure of the atmosphere. Scientists call instruments that measure air pressure barometers. In a U-tube barometer, one end is capped, and the other is open to the atmosphere. When the air pressure in the atmosphere increases, the water level on the open side moves down. When the air pressure in the atmosphere decreases, the water level on the open side moves up. In the atmosphere, the number density of the air decreases with increasing altitude. This causes the air pressure to decrease with increasing altitude.

Wrap-up

Key Question: How is the number of gas molecules in a sample related to pressure?

- The composition of Earth's atmosphere is not uniform. The density of the gas molecules in the air decreases with increasing altitude. This causes the pressure of the atmosphere to decrease with increasing altitude.
- The number density of a gas is the number of gas molecules per unit of volume, n/V.
- The pressure of a gas is directly proportional to the number of gas molecules per unit of volume. This relationship is written as $P = k \cdot \left(\frac{n}{V}\right)$.

Evaluate (5 minutes)

> **Check-in**
>
> A balloon is filled with helium, tied off, and then released. As it climbs into the air, its volume slowly increases. Explain what is going on with the helium atoms inside the balloon and the air molecules outside the balloon in terms of number density and pressure.

Answer: As the balloon ascends, the number density of the air molecules outside the balloon decreases, so the outside air pressure pushing in on the balloon decreases. The balloon expands until the pressure inside matches the pressure outside. The number of helium atoms does not change, but the number density inside the balloon decreases because the volume increases.

Homework

Assign the reading and exercises for Weather Lesson 15 in the student text.

LESSON 16 OVERVIEW

STP
The Mole and Avogadro's Law

Lesson Type
Classwork:
 Individuals
Demo:
 Individuals

Key Ideas

Avogadro's law states that equal volumes of gases contain equal numbers of gas molecules if they are at the same temperature and pressure. At a standard pressure of 1 atm and a standard temperature of 273 K, 22.4 L of *any* gas will contain exactly one mole of gas molecules. One mole represents 602,000,000,000,000,000,000,000 molecules.

As a result of this lesson, students will be able to

- define a mole
- explain Avogadro's law
- define standard temperature and pressure

Focus on Understanding

- The curriculum introduces the mole without overemphasizing the exact number the mole represents. At this point it is sufficient for students to understand that one mole represents an extraordinarily large number of atoms or molecules. The mole is covered more thoroughly in the next unit, Toxins.

- Students have a better grasp of the mole if it is not presented right away in scientific notation. They will use Avogadro's number and scientific notation extensively in the next unit, Toxins.

Key Terms

mole
standard temperature and pressure (STP)
Avogadro's law
Avogadro's number

What Takes Place

Students observe a simple demonstration with balloons filled with different gases. Students then complete a worksheet that expands on the concept of number density as it relates to gas pressure. Students are introduced to the notion that equal volumes of gas at the same temperature and pressure have equal numbers of gas molecules. Avogadro's hypothesis, STP, and the unit of a mole are introduced.

Materials

- student worksheet

Demonstration materials
- helium balloon
- balloon filled with CO_2 (your breath is fine)
- marking pen for labeling balloons
- box with a volume of 22.4 L (about 12 in. by 12 in. by 12 in.)

Setup

The demonstration requires that you have two balloons of the same volume containing different gases. Bring a helium balloon to class. Blow up a second balloon so that it matches the helium balloon in size. This is your CO_2 balloon. It is important that the two balloons look as if they have identical volumes. Clearly label each balloon with its identity.

LESSON 16 GUIDE

STP
The Mole and Avogadro's Law

Engage (5 minutes)

Key Question: How do chemists keep track of the number of gas particles?

> **ChemCatalyst**
>
> There are two balloons. One is filled with helium, He, and the other with carbon dioxide, CO_2.
>
> 1. Describe what happens when the balloons are released.
> 2. For the two balloons, state whether these properties are the same or different, and explain why:
>   - pressure, P
>   - temperature, T
>   - volume, V
>   - mass, m
>   - number density, n/V
>   - number, n
>   - density, m/V

Sample Answers: 1. The He balloon rises, while the CO_2 balloon falls. 2. All quantities except m and m/V are probably the same. P and T are the same because the pressure and temperature in the room are the same. The same pressure means the same number density, n/V. The volumes appear similar. If V and n/V are the same, then n must be the same. Because helium floats and carbon dioxide sinks, the densities and hence the masses must be different.

Discuss the ChemCatalyst

➡ Discuss the two balloons.

Sample Questions

- Why do the two balloons have the same number of particles?
- How can two gases have the same number of particles and the same volume but different densities?
- What do you think accounts for the different floating and falling behavior?
- Explain how you could measure out equal numbers of helium atoms and carbon dioxide molecules.

Explore (15 minutes)

Introduce the Classwork

➡ Write the following on the board and introduce the unit of a mole:

$$1 \text{ mole} = 602{,}000{,}000{,}000{,}000{,}000{,}000{,}000$$

140 *Living By Chemistry Teacher Guide* Unit 3 Weather

- Tell students that gas particles are far too small and too numerous to count individually. So a special counting unit is used, called a mole. One mole represents a very large number of gas particles, equal to 602,000,000,000,000,000,000,000, or 602 sextillion, particles.
- Explain that chemists use a unit called a *mole* to describe the number of gas particles in a sample. A gas sample may have 2 moles of gas or half a mole of gas or 0.1 mole of gas. Keep in mind that these all refer to very large numbers. For example, 2 moles is 2 times the number on the board. Half a mole is half that number.

➡ Pass out worksheets. Students will work individually on the worksheet to explore the relationship between gas pressure, P, and number of molecules, n.

LESSON 16 CLASSWORK
STP
The Mole and Avogadro's Law

Name _____
Date _____ Period _____

Purpose
To explore further the relationship between gas pressure and the number of gas particles in a sample.

Part 1: STP
Four samples of air have been collected at each of three different locations. Each air sample is in a different-size box. Assume that the temperature is 273 K at all three locations.

1. Fill in the missing amounts of moles of gas molecules for each box.

Volume (L)	Boxes filled at sea level P = 1 atm	Boxes filled on Mt. Denali P = 0.5 atm	Boxes filled outside airplane in flight P = 0.25 atm
11.2 L	0.500 mole	0.250 mole	0.125 mole
22.4 L	1.00 mole	0.500 mole	0.250 mole
33.6 L	1.50 moles	0.750 mole	0.375 mole
44.8 L	2.00 moles	1.00 mole	0.500 mole

2. Describe at least three patterns that you notice in the data.

   Sample answers: As the volume of the box increases, the number of moles of gas increases. As the pressure decreases, the number of moles of gas decreases. Both pressure and number of moles decrease with altitude. If you increase the size of the box, the number of moles will stay the same if the pressure decreases.

3. Analyze the data at sea level.
   a. What is the number density, n/V, in moles of gas molecules per liter of gas in the 22.4 L box at sea level?

   $n/V = 1.00$ mole$/22.4$ L $= 0.045$ mole/L

   b. What is the number density of the gas in all four air samples at sea level?

   $$\frac{0.500}{11.2} = \frac{1.00}{22.4} = \frac{1.50}{33.6} = \frac{2.00}{44.8} = 0.045 \text{ mole/L}$$

4. What is the number density for each location?

   n/V for sea level is 0.045 mole/L; n/V for Mt. Denali is 0.022 mole/L; n/V for outside the airplane is 0.011 mole/L.

5. Explain why the pressure is higher at sea level than on Mt. Denali.

   There are more moles of gas per liter at sea level. So there are more collisions and higher air pressure.

6. How many moles of gas would be in a 25.0 L box at sea level?

$n = kV$, where $k = n/V = 0.045$ mole/L for sea level.
(0.045 mole/L) · (25.0 L), or 1.13 moles of gas.

Part 2: Number and Mass

Consider samples of different gases. For each sample $T = 273$ K and $P = 1$ atm. (*Note:* The drawings simply represent the number of gas particles in correct proportion to one another.)

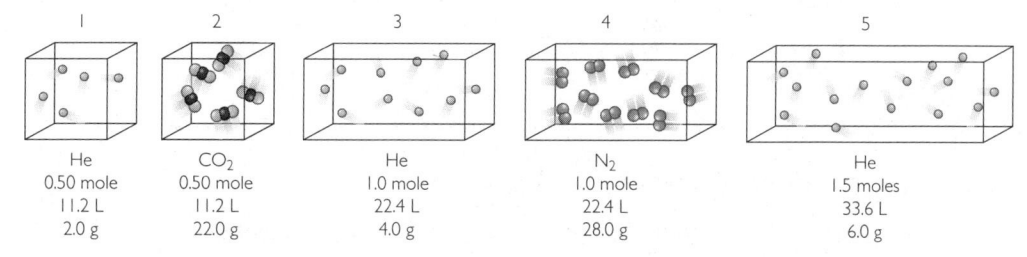

1	2	3	4	5
He	CO_2	He	N_2	He
0.50 mole	0.50 mole	1.0 mole	1.0 mole	1.5 moles
11.2 L	11.2 L	22.4 L	22.4 L	33.6 L
2.0 g	22.0 g	4.0 g	28.0 g	6.0 g

1. In the boxes showing helium gas, how many moles does each sphere represent?

   0.10 mole

2. Which box(es) has/have the most gas particles?

   Box 5 has the most. The 15 gas particles represent 1.5 moles.

3. Which box(es) has/have the most total atoms?

   Box 4. There are 10 N_2 molecules shown, with 2 atoms each.

4. There are twice as many total atoms in box 4 as in box 3, yet both boxes are at the same pressure. Explain why.

   Both boxes contain the same number of gas particles (atoms for He, molecules for N_2) per unit volume.

5. The masses of boxes 3 and 4 are different. Explain why.

   The number of gas particles is the same, but the nitrogen has more mass than the helium.

6. Describe or sketch a box containing 8.0 g of He atoms at 1 atm pressure. Show the relative number of He atoms and the size of this box compared to the size of the boxes in the table.

   The box should be twice the size of box 3 and contain 20 atoms.

7. **Making Sense** If you know that two gas samples are at the same temperature, what do you need to know in order to determine which gas is at a greater pressure?

   You need to know the number of particles in the sample and the volume of the container. The gas sample with the greater number of molecules per unit volume, n/V, will have the greater pressure.

8. **If You Finish Early** Consult the illustrations in Part 2. At a temperature of 273 K and a pressure of 1 atm, what volume does 1.0 mole of a gas occupy? Does it matter what the gas is? Explain.

   1.0 mole of gas occupies 22.4 L. The identity of the gas does not matter.

Explain and Elaborate (20 minutes)

Introduce Standard Temperature and Pressure, STP

Key Point

Standard temperature and pressure, or STP, is 1 atm of pressure at a temperature of 273 K. In order to be consistent in comparing gases, scientists have agreed on a standard set of conditions under which gases could be measured and compared: 1 atm of pressure and a temperature of 273 K. Note that this is 0 °C.

> **Standard temperature and pressure, STP:** One atmosphere of pressure and a temperature of 273 Kelvin.

Introduce Avogadro's Law

➡ Optional: Use the two balloons as a visual aid to help get across Avogadro's law.

Sample Questions

- Assume that the gases in the two balloons occupy the same volume and are at the same temperature and pressure. What else must be true about the two gas samples?
- Does it make sense that the numbers of gas particles in each of the two balloons are the same?
- How can two balloons hold identical numbers of gases but have different weights?

Key Points

Equal volumes of gases contain equal numbers of gas particles if the temperature and pressure are the same. This is true regardless of what gas is sampled. This generalization is known as Avogadro's law. It was first proposed by Italian scientist Amedeo Avogadro in 1811. Thus, if two gases have the same temperature, pressure, and volume, they also have the same number of gas particles.

There are exactly 602,000,000,000,000,000,000,000 particles in 22.4 L at STP. This is true for *any* gas. Chemists call this enormous number 1 mole. The abbreviation for this unit is mol.

> **Mole:** A unit invented by chemists to count large numbers of gas particles. There are 602,000,000,000,000,000,000,000 particles in 1 mole. This is 602 sextillion.

Wrap-up

Key Question: How do chemists keep track of the number of gas particles?

- Avogadro's law states that equal volumes of gases contain the same number of particles if they are at the same temperature and pressure. This holds true for all gases.
- Gases are often compared at a standard temperature and pressure of 1 atm and 273 K. This is also referred to as STP.

- At STP, any *gas* will occupy 22.4 L and consist of exactly 602 sextillion, or 602,000,000,000,000,000,000,000, particles.

Evaluate (5 minutes)

> **Check-in**
>
> One balloon contains 22.4 L of Ar, argon gas, and another balloon contains 22.4 L of Ne, neon gas. Both balloons are at 273 K and 1 atm.
>
> 1. Do the balloons contain the same number of atoms? Why or why not?
> 2. Will the balloons have the same mass? Why or why not?

Answers: 1. The balloons contain the same number of atoms because they are at the same temperature and pressure, STP. 2. The balloons will not have the same mass. Argon atoms are heavier than neon atoms.

Homework

Assign the reading and exercises for Weather Lesson 16 in the student text.

LESSON 17 OVERVIEW

Take a Breath
Ideal Gas Law

Lesson Type
Lab:
 Groups of 4

Key Ideas

The ideal gas law, $PV = nRT$, allows scientists to relate gas pressure, volume, moles of particles, and temperature. R is a constant and does not change; however, its value depends on the units that are used. If you know pressure, volume, and temperature, you can calculate the number of moles of gas in a sample by using this equation.

As a result of this lesson, students will be able to

- define the ideal gas law
- define the universal gas constant, R
- complete calculations for finding n, using the ideal gas law

Focus on Understanding

- Note that we still are not using scientific notation for the number of molecules in a mole. This allows students to become familiar with this number and grasp its magnitude.

Key Terms

ideal gas law
universal gas constant (R)

What Takes Place

SciLinks
Topic: Gas Laws
Visit: www.SciLinks.org
Web code: KEY-306

Students determine the volume of one average breath of air from their lungs. They blow air through a tube into a 2 L plastic soft drink bottle, displacing an amount of water. Students use their measurements to calculate the volume of a single breath. They then use the ideal gas law to calculate the number of moles of air in one breath at sea level and the number of moles of air in one breath at a higher elevation.

Materials

- student worksheet

Per group of 4

- 2 L plastic soft drink bottle with cap
- sink or other large container for water (at least 5 L)
- tap water
- 3 ft of flexible tubing (U-tube from Weather Lesson 15: *n* Is for Number Density)

- 4 drinking straws
- watercolor marker or overhead pen
- 250 mL or 500 mL graduated cylinder

Setup

Cut drinking straws into three or four pieces to serve as mouthpieces for the tubing. Each student should use a different straw. You might want to have your own equipment set up at the front of the room to demonstrate the procedure.

LESSON 17 GUIDE

Take a Breath
Ideal Gas Law

Engage (10 minutes)

Key Question: How can you calculate the number of moles of a gas if you know P, V, and T?

> **ChemCatalyst**
> 1. Describe how you can determine the volume of a breath of air.
> 2. Name four factors that might affect the volume you measure.
> 3. What do you need to know in order to determine the number of molecules in a breath of air?

Sample Answers: 1. You could blow into a bag and measure the volume of the bag. 2. The volume you measure will depend on how deep a breath you take, how large a person you are, the pressure in the atmosphere (or your altitude), and the air temperature. 3. You need to know the relationship between the number of molecules and volume.

Discuss the ChemCatalyst

➡ Discuss measuring a breath of air.

Sample Questions

- What type of container would you use to measure the volume of a breath? (a flexible container)
- How large a container do you think you need? (several liters)
- Do you think it is sufficient to take only one measurement? Why or why not?
- What variables do you think affect the number of molecules in a breath of air? (the volume of a breath, air pressure, air temperature, your size, how deeply you breathe)

Explore (20 minutes)

Introduce the Lab

➡ Tell students that the ideal gas law allows us to determine the number of moles of gas molecules in a gas sample if temperature, pressure, and volume are known. The equation for the ideal gas law is

$$PV = nRT$$

where R is equivalent to the proportionality constant, k, for this equation.

$$R = \frac{PV}{nT} \qquad R = 0.082 \text{ L} \cdot \text{atm/mol} \cdot \text{K}$$

148 Living By Chemistry Teacher Guide   Unit 3 Weather

- Ask students to work in groups of four. Let them know they will need their safety goggles.
- Go over the general procedure for Part 1 with the class. Emphasize to students that the goal is to find the volume of a normal breath. (This is not a competition to determine who can exhale the largest volume of air.)
- In Part 2, students complete calculations using the volume of one breath and the ideal gas law to determine how many moles of air particles they would breathe at different altitudes.

LESSON 17 LAB

Take a Breath
Ideal Gas Law

Name _____
Date _____ Period _____

Purpose
To determine the number of moles of air particles in an average breath and to explore the ideal gas law.

Materials
- 2 L plastic soft drink bottle with cap
- sink or other large container
- tap water
- 3 ft of flexible tubing
- 4 drinking straws
- marker
- 250 mL or 500 mL graduated cylinder

Part 1: Volume of a Breath of Air
The goal of this part of the lab is to determine the volume of one normal breath of air for each person in your group. The outline for a procedure is given below. Your group will need to decide how you will figure out the air volume.

Procedure
1. Fill a larger container or tub about half full with tap water.
2. Fill a 2 L plastic soft drink bottle with tap water. Put the cap on loosely.
3. Carefully turn the bottle upside down without spilling any water.
4. Put the bottle into the large container of water so that the mouth of the bottle is underwater. Remove the cap underwater.
5. Feed the flexible tubing under the water so that one end goes inside the bottle.
6. Put your straw into the other end of the tubing. Do not share straws.
7. When it is your turn, exhale into the straw to collect the air of one normal breath.
8. With a marking pen, mark the volume of air on the soft drink bottle.
9. Figure out the volume of the air trapped inside the bottle. Record this volume.
10. Repeat the procedure for each person in the group. Replace the straw for each new person.

Part 2: Analysis
Use the volume you determined for one breath for your calculations in Questions 1 and 2. Also use the ideal gas law described by $PV = nRT$ where $R = 0.082$ L · atm/mol · K.

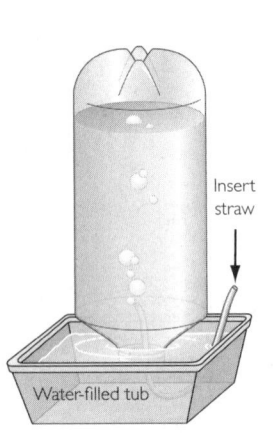

1. **Moles in a Breath at Sea Level.** Suppose you take a breath at sea level where the air pressure is 1.0 atm and the temperature is 25 °C. Use the ideal gas law to determine the number of moles of air molecules in one breath.

   Sample answer: Breath volume = 0.50 L
   Convert Celsius to Kelvin: 25 °C = 298 K
   $PV = nRT$
   $(1.0 \text{ atm}) \cdot (0.50 \text{ L}) = n \left(0.082 \frac{L \cdot atm}{mole \cdot K}\right)(298 \text{ K})$
   n = 0.020 mol

2. **Moles in a Breath on a Mountaintop.** Suppose you take a breath on a mountaintop at altitude 10,000 ft where the air pressure is 0.75 atm and the temperature is 20 °C. Use the ideal gas law to determine the number of moles of air molecules in one breath.

   $(0.75 \text{ atm}) \cdot (0.50 \text{ L}) = n \left(0.082 \frac{L \cdot atm}{mole \cdot K}\right)(293 \text{ K})$
   Solve for n: n = PV/RT = 0.016 mol

3. There are 602 sextillion, or 602,000,000,000,000,000,000,000, gas molecules in 1 mole. Calculate the number of gas molecules in a breath at sea level. Calculate the number of gas molecules in a breath on a mountaintop. (Show your work.)

   Sample answer at sea level:
   (0.020)(602,000,000,000,000,000,000,000) =
   12,000,000,000,000,000,000,000 molecules
   Sample answer at 10,000 ft:
   (0.016)(602,000,000,000,000,000,000,000) =
   9,600,000,000,000,000,000,000 molecules

4. Based on your calculations in Question 3, what is the difference between the number of molecules in one normal breath at sea level and the number at 10,000 ft?

   There are about 3 sextillion fewer air molecules at 10,000 ft.

5. The air in airplanes is "pressurized." What do you think this means?

   Air must be pumped into airplanes so that people get enough oxygen to breathe. This means that the air pressure inside the plane is greater than the air pressure outside the plane.

6. **Making Sense** Explain what the ideal gas law can help you figure out. What must you know before you can use the ideal gas law?

   The ideal gas law can assist in figuring out how many moles of gas particles are present in a sample of gas under certain conditions. You must know the pressure, volume, and temperature of the gas and the proportionality constant, R.

7. **If You Finish Early** Use the ideal gas law and your breath volume to figure out how many moles of air molecules would be in one breath at the top of Mount Everest. Air pressure at 29,000 ft is 0.30 atm. The temperature at the summit at the warmest time of the year is −19 °C.

   $(0.30 \text{ atm})(0.50 \text{ L}) = n \left(0.082 \frac{L \cdot atm}{mol \cdot K}\right)(254 \text{ K})$
   $(0.15) = n(20.83)$
   n = 0.007 mol

Explain and Elaborate (15 minutes)

Discuss How Groups Determined the Volume of One Breath of Air

➡ You might have each group write its results for the volume of one breath of air on the board. (Note that the average total lung capacity for an adult is around 6 L of air. However, the amount of air breathed in or out during normal respiration is closer to 500 mL.)

Sample Questions

- What are some different ways you could determine the volume of one breath of air?
- What are some reasons why the calculated volumes differ from group to group?
- What are some possible causes of error in this procedure?
- How could you improve the experiment you did? (measure the volume of one breath several times, measure several breaths and divide by the number of breaths, collect the breath in a larger bottle, try to breathe normally, etc.)

Key Points

We should see differences in the volume of one breath of air from group to group. Each person's lung capacity is slightly different. In addition, some students might take a deep breath, while others might take a shallow breath. Because there is variation from one breath to the next, it may be useful to measure the volumes of several breaths and average the outcomes.

There is more than one way to figure out the volume of air that was exhaled into the bottle.

- One way is to begin the procedure by calibrating the bottle. You can do this by putting known quantities of water (say, 250 mL or 500 mL) into the bottle and marking the side with a marker.
- Another method for measuring breath volume is to (1) breathe into the bottle, (2) cap off the bottle underwater, (3) remove the bottle from the tub and invert it, and (4) refill the bottle with water using measured amounts.
- A third method for figuring out the breath volume is to measure the amount of water that is left behind in the 2 L bottle and then subtract that amount from 2 L. This gives you the amount of volume occupied by the air. It might be more precise to measure an exact volume of the bottle before starting, in case it is slightly over or under 2 L.
- A fourth method is to keep breathing into the bottle until all the water is displaced. Then you can divide the total volume of the bottle by the number of breaths.

Discuss the Ideal Gas Law

➡ You might want to guide the class through the last question on the worksheet. Ask student volunteers to assist you at the board.

Sample Questions

- How did you use the ideal gas law to determine the number of moles of gas particles in one breath?

- How did you convert moles to number of gas particles?
- What did you find out about a breath of air at sea level compared to a breath of air on a 10,000-ft-high mountain?
- Describe how you would figure out the number of moles of air in a breath at the top of Mount Everest, or at 30,000 ft elevation.

Key Points

The ideal gas law allows scientists to relate gas pressure, volume, moles of particles, and temperature. The ideal gas law is a result of combining the ideas in the combined gas law with Avogadro's law. The ideal gas law allows you to calculate the number of moles of gas particles in any given volume. You must know the pressure and temperature of the gas, and the values must be expressed in the appropriate units (in this case, pressure in atmospheres, volume in liters, temperature in Kelvin, and number in moles).

Ideal Gas Law

The ideal gas law states that $PV = nRT$, where R, the universal gas constant, is equivalent to the proportionality constant, k, for this equation.

$$R = \frac{PV}{nT} \qquad R = 0.082 \text{ L} \cdot \text{atm/mol} \cdot \text{K}$$

Note that R is the same for all gases but the value of R does change depending on if the units change. For example, $R = 62.4$ L · mmHg/mol · K and $R = 8.314$ L · kPa/mol · K. We have opted to round off R to 3 significant digits for simplicity. Note that the value for R is often reported to more significant digits than this.

In today's activity the results for the value of n vary from group to group. However, if a breath volume is 0.50 L, then n is equal to 0.020 mole at sea level and 0.016 mole at 10,000 ft.

The number of moles can be converted to the total number of gas molecules by multiplying by 602 sextillion. The difference in number of moles between sea level and 10,000 ft is about 0.004 mole. This small difference in moles is about 2.4 sextillion total molecules. So there are many fewer molecules in each breath at 10,000 ft than at sea level.

The ideal gas law can be used to solve for other variables besides n. For example, you can use the ideal gas law to solve for P if n, V, and T are known.

Wrap-up

Key Question: How can you calculate the number of moles of a gas if you know P, V, and T?

- The ideal gas law relates volume, pressure, temperature, and the number of moles of a gas sample: $PV = nRT$, where $R = 0.082$ L · atm/mol · K.
- R is a number that relates all the different units to one another in the ideal gas law. Its value does not change if the units don't change. R is called the universal gas constant.
- The ideal gas law can be used to figure out P, V, n, or T when the other three variables are known.

Evaluate (5 minutes)

Check-in

You cap a 1.0 L plastic bottle on a mountaintop where the air pressure is 0.50 atm and the temperature is 298 K.

1. How many moles of gas are in the bottle?
2. What is the number density, n/V, of the gas inside the bottle on the mountaintop?
3. At sea level, the volume of the bottle becomes 0.50 L. What is the number density of the gas inside the bottle at sea level?

Answers: 1. Using the ideal gas law to solve for n: $n = PV/RT = 0.020$ mol. 2. Number density on the mountaintop: $n/V = 0.020$ mol/1.0 L $= 0.020$ mol/L. 3. Number density at sea level: $n/V = 0.020$ mol/0.50 L $= 0.040$ mol/L.

Homework

Assign the reading and exercises for Weather Lesson 17 in the student text.

LESSON 18 OVERVIEW

Feeling Humid
Humidity, Condensation

Lesson Type
Lab:
Groups of 4

Key Ideas

Humidity is the amount of water vapor in a volume of air. Thus, it is a measure of water vapor density, n/V. Humidity can be expressed as mass density in grams per cubic centimeter or as number density, in moles per 1000 liters. Air temperature affects how much water vapor potentially can be in the air. Humidity generally is expressed as relative humidity, or the percentage of water vapor in the air relative to the maximum amount possible at that temperature. At maximum humidity, the air is saturated with water. Maximum humidity increases with increasing air temperature.

As a result of this lesson, students will be able to

- define humidity and relative humidity
- explain the relationship between humidity and phenomena such as cloud formation, fog, rainfall, and dew
- explain the relationship between water vapor density and air temperature

Focus on Understanding

- What meteorologists refer to as humidity is not water vapor density but actually relative humidity, which is expressed as a percent.

Key Terms

humidity
relative humidity

What Takes Place

Students investigate the amount of water vapor in the air. They also explore the relationships among condensation, humidity, and air temperature. They complete two different procedures that allow them to determine the humidity of the air in their classroom. One procedure involves discovering the temperature at which condensation occurs. The other involves a wet-bulb thermometer. If you are pressed for time, Part II is optional or can be done as a demo.

Materials

- student worksheet
- transparencies—Water Vapor Density Versus Temperature graph
- Handout—Relative Humidity (optional)

Section III Lesson 18 Feeling Humid

Per group of 4
- room-temperature water
- 250 mL beaker
- 2 thermometers
- spray bottle with water
- rubber bands, tiny
- cheesecloth (~2 in. by 6 in.)
- ice cubes
- stirring rod

Setup

You can set up half the stations for Part 1 and half for Part 2. For Part 1, set up a 250 mL beaker, water, ice, thermometer, and stirring rod. For Part 2, set up a beaker containing 200 mL of room-temperature water, two thermometers, small rubber bands, and a small piece of cheesecloth—approximately 2 in. by 6 in.—for students to wrap around the bulb of the thermometer.

LESSON 18 GUIDE

Feeling Humid
Humidity, Condensation

Engage (5 minutes)

Key Question: What is humidity and how is it measured?

> **ChemCatalyst**
>
> 1. Is there water vapor in the air right now? What evidence do you have to support your answer?
> 2. What do you think humidity means? How does humidity depend on temperature?

Sample Answers: 1. Most students will say that some water vapor is in the air at all times because water is evaporating all around us or that clouds can be seen overhead. 2. Humidity is a measure of the amount of water vapor in the air. Students will probably say that warm air can "hold" more water vapor than cold air.

Discuss the ChemCatalyst

➡ Assist the class in sharing ideas about humidity.

Sample Questions

- Does the amount of water vapor in the air change? Explain.
- What do you think affects the amount of water vapor in the air?
- How do meteorologists determine the moisture content of the air? How do they express how much water vapor is in the air?
- What is humidity? Do you feel warmer or colder when there is humidity?

Explore (15 minutes)
Introduce the Lab

➡ Introduce the term *humidity*. Tell students that humidity describes the amount of water vapor in the air.

> **Humidity:** The density of the water vapor in the air at any given time. Humidity is dependent on air temperature and pressure.

➡ Pass out worksheets. Explain to students that they will be working in groups of four to complete two procedures to determine the amount of water vapor in the air.

Section III　　　　　　　　　　　　　　　　　　　　　Lesson 18 Feeling Humid　　157

LESSON 18 LAB

Feeling Humid
Humidity, Condensation

Name _____
Date _____ Period _____

Purpose
To explore water vapor density and determine the amount of water vapor in the air.

Materials
- room-temperature water
- 250 mL beaker
- 2 thermometers
- spray bottle with water
- rubber bands
- cheesecloth or paper towel
- ice cubes
- stirring rod

Part 1: Condensation on Glass
Procedure

1. Put about 150 mL of room-temperature water into a 250 mL beaker. Record the water temperature. Look for condensation, or moisture on the outside of the glass. Make a table like the one here to record your data.

Time	Temperature (°C)	Condensation?
Before ice		
0 s (ice added)		
30 s		

2. Add ice to the water until the beaker is close to full. Start timing. Stir, and record the temperature again. Continue to look for condensation.
3. Continue stirring and recording the temperature every 30 sec. Note the temperature at which condensation first forms on the outside of the beaker. Record this temperature and stop.

Analysis

1. At what temperature did water form on the outside of the beaker?

    Answers will vary with weather conditions.

2. Where did the water come from that ended up on the outside of the glass?

    The water came from the atmosphere.

3. Imagine that the water vapor density of the air in your lab increases significantly. What effect do you think this would have on the formation of water on the beaker?

    Water would condense on the beaker sooner, at a higher temperature.

Living By Chemistry Teaching and Classroom Masters: Units 1–3
© 2010 Key Curriculum Press

4. Imagine repeating the experiment on a day when the temperature in the lab is 10 °C cooler. What effect would this have on the formation of the water on the beaker?

If the air is cooler, there may be less moisture in the air. Condensation may not show up until the beaker is colder.

There is a limit to the amount of water vapor that can be present in the air. This limit depends on the air temperature. The graph shows the maximum water vapor density in the atmosphere at different temperatures, expressed in mol/1000 L.

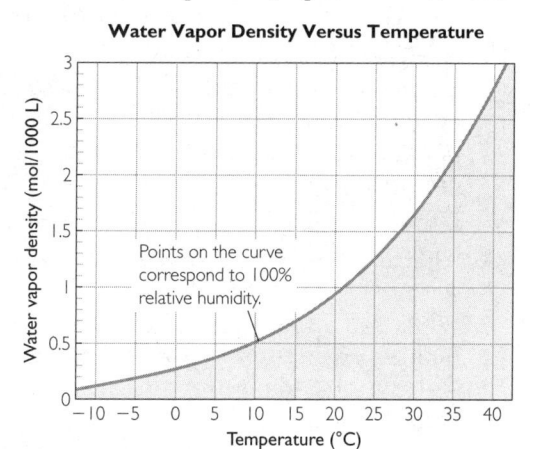

Water Vapor Density Versus Temperature

Points on the curve correspond to 100% relative humidity.

5. On the x-axis of the graph, find the condensation temperature you recorded in your data. What is the water vapor density of the air at this temperature in moles/1000 L?

At 3 °C the water vapor density is somewhere around 0.35 mol of water per 1000 L of air sampled.

6. What is the maximum water vapor density possible at a temperature of 40 °C?

2.85 mol per 1000 L.

7. At what temperature is the maximum water vapor density equal to 0.71 mol per 1000 L? Explain in simple language what this number means.

At approximately 15 °C. This number means that at 15 °C the maximum amount of water vapor that can be in the air without condensing out is 0.71 mol of water vapor for every 1000 L of air.

8. Is it possible for the water vapor density in the air to reach 1.50 moles/1000 L at a temperature of 25 °C? Explain why or why not.

No. The maximum amount of water vapor possible at 25 °C is 1.29 mol/1000 L. Any additional water will remain liquid.

Part 2: Wet-Bulb Thermometer

Procedure and Questions

1. Record the temperature of the room with the dry thermometer.
2. Dip the cheesecloth or paper towel in water and wrap it around the bottom of the second thermometer. If necessary, use a rubber band to hold it in place.

3. Securely holding the middle of each thermometer, carefully wave them both in the air for 2 minutes. Record the temperature of each thermometer.

Dry-bulb temperature	Wet-bulb temperature	Difference
20 °C	14 °C	6°

1. What happened to the temperature of the thermometer with the cheesecloth wrapped around it? How can you explain the change in temperature?

   Answers will vary. Waving the thermometer with the cheesecloth around in the air made more water evaporate. When water evaporates, it cools off whatever it is touching.

2. Use the Handout—Relative Humidity to determine the humidity of the air in your classroom.

   In this example, the room would be at 51% relative humidity.

3. When the difference in bulb temperatures is smaller, is the relative humidity higher or lower? How can you explain this?

   When the difference in bulb temperatures is smaller, the relative humidity is higher. Less water can evaporate when a lot of water is already in the air.

4. What would you notice about the difference between the wet-bulb temperature and the dry-bulb temperature if the air in the room contained very little moisture?

   As humidity decreases, the difference between the wet- and dry-bulb readings increases.

5. **Making Sense** What is water vapor density? What does it mean that water vapor density reaches a maximum at a specified temperature?

   Water vapor density is the number of water molecules that are in gaseous form per unit of volume of air. When the maximum water vapor density is reached, the greatest possible amount of water has evaporated into the air for a given temperature.

6. **If You Finish Early** When the water vapor density reaches the maximum amount at a given temperature, the humidity is considered to be 100%. Under which conditions is the water vapor density greater—50% humidity at 10 °C, or 25% humidity at 25 °C? Explain your thinking.

   A humidity of 25% at 25 °C puts more water vapor into the air than does 50% humidity at 10 °C. If you look at the graph, half of the amount of water vapor (50% humidity) that can go into the air at 10 °C is about 0.26 mol per 1000 L. At 25 °C, 100% humidity places about 1.30 mol per 1000 L into the air. At 25% humidity, there is 0.33 mol per 1000 L.

RELATIVE HUMIDITY

Relative Humidity (%)

Dry-Bulb Temperature (°C)	\multicolumn{16}{c}{Difference Between Wet-Bulb and Dry-Bulb Temperatures (°C)}
	0	1	2	3	4	5	6	7	8	9	10	11	12	13	14	15
−20	100	28														
−18	100	40														
−16	100	48														
−14	100	55	11													
−12	100	61	23													
−10	100	66	33													
−8	100	71	41	13												
−6	100	73	48	20												
−4	100	77	54	32	11											
−2	100	79	58	37	20	1										
0	100	81	63	45	28	11										
2	100	83	67	51	36	20	6									
4	100	85	70	56	42	27	14									
6	100	86	72	59	46	35	22	10								
8	100	87	74	62	51	39	26	17	6							
10	100	88	76	65	54	43	33	24	13	4						
12	100	88	78	67	57	48	38	28	19	10	2					
14	100	89	79	69	60	50	41	33	25	16	8	1				
16	100	90	80	71	62	54	45	37	29	21	14	7	1			
18	100	91	81	72	64	56	48	40	33	26	19	12	6			
20	100	91	82	74	66	58	51	44	36	30	23	17	11	5		
22	100	92	83	75	68	60	53	46	40	33	27	21	15	10	4	
24	100	92	84	76	69	62	55	49	42	36	30	25	20	14	9	4
26	100	92	85	77	70	64	57	51	45	39	34	28	23	18	13	9
28	100	93	86	78	71	65	59	53	47	42	36	31	26	21	17	12
30	100	93	86	79	72	66	61	55	49	44	39	34	29	25	20	16

254 Unit 3 Weather
Lesson 18 • Handout

Living By Chemistry Teaching and Classroom Masters: Units 1–3
© 2010 Key Curriculum Press

Explain and Elaborate (20 minutes)
Discuss Water Vapor Density and Humidity

Sample Questions

- What does water vapor density describe?
- What does the condensation procedure indicate?
- Under what conditions would the water vapor condense out of the air at a lower temperature? at a higher temperature?
- What is the water vapor density of the air in your classroom? Explain how you know.

Key Points

Humidity is a measure of the amount of water vapor in the air. Humidity can be expressed as number density, n/V, or as mass density, g/cm^3. Temperature affects maximum humidity. Warm air can have a higher maximum water vapor density than cold air.

The condensation procedure provides evidence that water vapor is present in the air. When the temperature drops low enough, the water vapor in contact with the beaker changes phase and condenses on the outside of the glass. There is no other place this moisture could come from. Water will condense from air if the air cools sufficiently.

The temperature at which water vapor condenses indicates how much water vapor is in the air. Suppose water condenses at a temperature of 15 °C in your classroom. The graph shows that this is equivalent to approximately 0.70 mol of water vapor per 1000 L of air.

Introduce Relative Humidity

[T] ➡ Display the transparency Water Vapor Density Versus Temperature.

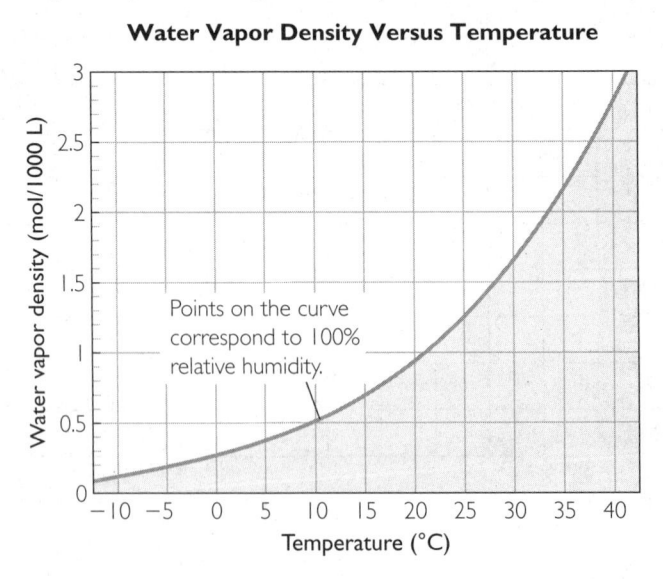

Water Vapor Density Versus Temperature

[Graph: Water vapor density (mol/1000 L) on y-axis from 0 to 3, Temperature (°C) on x-axis from −10 to 40. Curve rises exponentially. Annotation: "Points on the curve correspond to 100% relative humidity."]

Sample Questions

- What does the curve on the graph of water vapor density versus temperature represent? What do points in the shaded area represent?

- Is it possible for there to be any values above the curve, say 3 moles of water vapor per 1000 L of air at 30 °C? Explain.
- Once the air has reached a maximum water vapor density for the temperature of the air, what happens?

Key Points

There is an upper limit to the amount of water vapor that can be present in the atmosphere at a given temperature. If you look at the graph, the maximum amount of water vapor that can be present at 30 °C is approximately 1.69 moles of water vapor per 1000 L of air sampled. The maximum water vapor density is called 100% relative humidity. At 100% relative humidity, the air is saturated with water vapor, and no further evaporation can take place. The water vapor density corresponding to 100% humidity depends on temperature. Thus, conditions of 100% humidity at 40 °C means more water vapor in the air than 100% humidity at 10 °C.

Humidity is sometimes expressed as relative humidity. Relative humidity is a percentage of the maximum humidity at a specified temperature. If the relative humidity is 50%, the amount of water vapor in the air is half the maximum amount possible for that temperature. At 90% humidity, you would expect it to be very close to raining or very foggy. You would almost feel the moisture in the air. Summers in the Midwest can have very high humidities. People tend to feel most comfortable at a relative humidity of about 45%.

> **Relative humidity:** The amount of water vapor in the air compared to the maximum amount of water vapor possible for a specific temperature, expressed as a percent.

Process the Wet- and Dry-Bulb Procedure (optional)

⟹ Use the handout showing relative humidity.

Sample Questions

- What did the wet- and dry-bulb procedure allow you to determine?
- Examine the table of relative humidity. What does a large temperature difference between wet and dry bulbs suggest? a small temperature difference?
- Suppose your classroom has a relative humidity of 60%. What does that mean?

Key Points

The wet- and dry-bulb procedure allows you to determine the relative humidity of the air. While the thermometer is being waved around, the water evaporates from the wick, cooling the wet-bulb thermometer. When water evaporates, it transfers heat away from whatever it is in contact with. That is why sweating cools your skin.

The larger the temperature difference between the wet and dry bulbs, the drier the air. Likewise, the smaller the temperature difference between the bulbs, the moister the air. If the surrounding air is dry, more moisture evaporates from the wick, cooling the wet-bulb thermometer more. This creates a greater difference between the temperatures of the two thermometers. If the surrounding air is at a relative humidity of 100%, there is no difference between the two temperatures.

Meteorologists have worked out charts of these differences for each degree of temperature so that observers can find relative humidity easily.

Wrap-up

Key Question: What is humidity and how is it measured?

- Humidity is a measure of the amount of water vapor in the air. It can be expressed as water vapor density or as relative humidity.
- Water vapor density is affected by both air temperature and air pressure.
- There is a limit to the amount of water vapor that can be present in the air at a given temperature. The maximum amount possible is called 100% relative humidity.

Evaluate (5 minutes)

> **Check-in**
>
> On a hot summer day, a firefighter records a dry-bulb temperature of 30 °C and a wet-bulb temperature of 12 °C. What does this tell you about the relative humidity?

Answer: The difference in the two bulb temperatures is 18°. This number is not even on the table, but it is apparent that the air is quite dry. There is very little water vapor in the air, perhaps as low as 4% relative humidity.

Homework

Assign the reading and exercises for Weather Lesson 18 in the student text.

Optional: Assign a project from the student text.

LESSON 19

Hurricane!
Extreme Physical Change

OVERVIEW

Lesson Type
Classwork: Individuals

Key Ideas

Hurricanes form as a result of specific weather conditions: very warm oceans, humid air, and converging winds. A continuing cycle of evaporation and condensation drives the hurricane by setting up a large air pressure differential. The gas-to-liquid phase change fuels hurricanes. This phase change is directly related to air and ocean temperatures. Concerns have been raised that global warming may be increasing the frequency and/or intensity of hurricanes.

As a result of this lesson, students will be able to

- describe the meteorological conditions that result in a hurricane
- explain the role of phase change, air pressure, and temperature in hurricane formation
- define climate and global warming

What Takes Place

Students complete a worksheet about the anatomy and physical characteristics of a hurricane. The conditions that affect the formation and intensity of hurricanes are also explored, with a focus on changes in temperature. The discussion and the reading in the student text make connections between hurricanes and global warming. If you are pressed for time, Part I is optional.

Materials

- student worksheet
- transparencies—ChemCatalyst, Global Warming Trends, Anatomy of a Hurricane, Water Vapor Density Versus Temperature graph (from Lesson 18)
- Handout—Data for 2005 Hurricane Season
- projection system for displaying satellite images (optional)

SciLINKS NSTA
Topic: Severe Weather
Visit: www.SciLinks.org
Web code: KEY-319

Setup

If you have the necessary technology in your classroom, a wealth of satellite images and footage of hurricanes is available from NASA and other organizations.

LESSON 19 GUIDE

Hurricane!
Extreme Physical Change

Engage (5 minutes)

Key Question: What are hurricanes and what causes them?

ChemCatalyst

1. What is a hurricane? What characteristics does it have?
2. Where do hurricanes form? What can you tell about a hurricane from the satellite image?

Sample Answers: 1. Hurricanes are large, spinning storms that bring high winds and lots of rain. When they move over landmasses, they can do tremendous damage. 2. Hurricanes form over the ocean. The satellite image shows a hurricane's characteristic spiral shape and its calm center, called the "eye." They spin counterclockwise in the Northern Hemisphere.

Discuss the ChemCatalyst

Sample Questions

- Where and when do hurricanes form?
- What makes a hurricane destructive?
- In what direction do hurricanes turn? What explains this observation?
- What is the difference between a hurricane and a large rainstorm? a tornado?
- What is the "eye" of the hurricane?

Explore (15 minutes)

Introduce the Classwork

➡ Pass out worksheets. Students can work individually on the worksheet.

166 *Living By Chemistry Teacher Guide* Unit 3 Weather

LESSON 19 CLASSWORK

Hurricane!
Extreme Physical Change

Name _____
Date _____ Period _____

Purpose
To learn about hurricanes and the variables that affect their formation and intensity.

Part 1: Hurricanes
The handout contains some data for the 2005 hurricane season. Use the handout to answer the questions.

1. What do you think is the difference between a tropical storm, a hurricane, and a tropical depression?

   According to this table, the wind speed and air pressure make the difference. Tropical depressions have wind speeds of around 35 mi/h. Tropical storms have wind speeds between 35 and 70 mi/h. Hurricane winds are over 70 mi/h. Hurricanes also have the lowest air pressure readings.

2. Which hurricanes were the least intense in 2005? What is your evidence?

   Hurricanes Ophelia and Cindy were the least intense in 2005. They are category 1 hurricanes, with less intense winds.

3. Which hurricanes were the most intense in 2005? What is your evidence?

   Hurricanes Katrina, Rita, and Wilma. They are all category 5 hurricanes, with wind speeds of about 175 mi/h.

4. What approximate air pressure range is associated with hurricanes?

   The air pressure ranges between 882 and 992 millibars.

5. What wind speed range is associated with hurricanes?

   The wind speed ranges from 75 to 175 mi/h.

6. How do you think the category number of a hurricane is determined?

   The category varies with the wind speed.

7. If a storm system has an air pressure of 980 mb, do you think it will be classified as a hurricane? What is your reasoning?

   It depends on the wind speed of the storm. The table shows a tropical storm with an air pressure of 980 mb, but its wind speed is only 70 mi/h.

8. Hurricanes form only in places where the ocean water is at least 80 °F. What effect does high temperature have on the water vapor density of the air over the ocean?

   More moisture is in the air at higher temperatures. The water vapor density is higher where it is warm.

9. When do you think hurricane season is for the East Coast of North America?

   July through October.

Part 2: Hurricanes and Temperature

1. As a hurricane travels across the surface of the ocean, its wind speed changes with the temperature of the water. This graph shows wind speed versus ocean temperature for a hurricane.

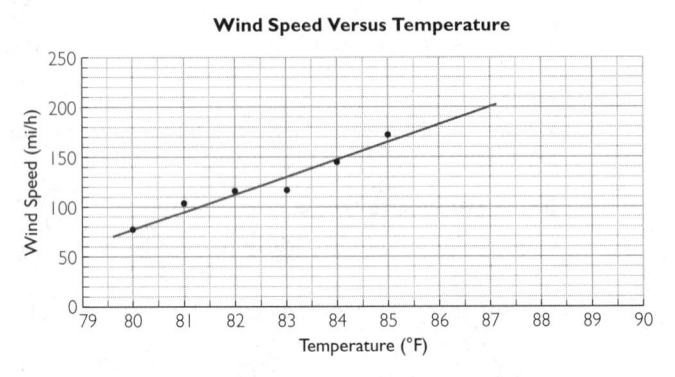

Wind Speed Versus Temperature

2. How does the wind speed change with ocean surface temperature?

   The wind speed increases with higher ocean surface temperatures.

3. If the planet warms 2 °F in the next 30 years, what wind speeds can you predict for the most severe hurricanes at that time?

   Wind speeds of approximately 190–200 mi/h.

4. This graph shows wind speed versus pressure for six hurricanes. What pressure corresponds to the wind speed you determined in Question 3?

   If the answer to Question 3 is about 200 mi/h, then the air pressure extrapolated from the graph is about 840 mb.

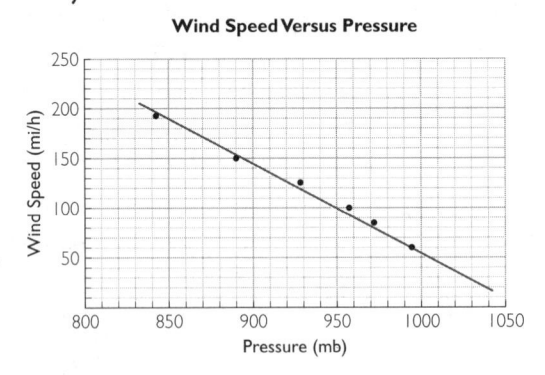

   Wind Speed Versus Pressure

5. **Making Sense** What factors affect the severity of a hurricane?

   The temperature of the ocean water affects the severity of a hurricane—the warmer the water, the higher the wind speed. The air pressure affects the severity of a hurricane—the lower the air pressure, the higher the wind speed.

DATA FOR 2005 HURRICANE SEASON

Name	Date	Wind speed (mi/h)	Pressure (millibars)	Category
Tropical storm Bret	June 28–30	40	1002	—
Hurricane Cindy	July 3–7	75	992	1
Hurricane Dennis	July 5–13	150	930	4
Hurricane Emily	July 11–21	155	929	4
Tropical storm Franklin	July 21–29	70	997	—
Tropical storm Harvey	Aug 2–8	63	994	—
Hurricane Irene	Aug 4–18	98	975	2
Tropical depression 10	Aug 13–14	35	1008	—
Hurricane Katrina	Aug 23–31	173	902	5
Tropical storm Lee	Aug 28–Sep 2	40	1007	—
Hurricane Maria	Sep 1–10	115	960	3
Hurricane Ophelia	Sep 6–18	92	976	1
Hurricane Rita	Sep 18–26	173	897	5
Tropical depression 19	Sep 30–Oct 2	35	1006	—
Hurricane Vince	Oct 9–11	75	987	1
Hurricane Wilma	Oct 15–25	175	882	5
Tropical storm Delta	Nov 23–28	70	980	—

Explain and Elaborate (20 minutes)
Introduce Hurricanes and Their General Characteristics

Sample Questions

- What is a hurricane?
- How are hurricanes ranked?

Key Points

Hurricanes are destructive storms characterized by strong winds and large amounts of rainfall. Hurricanes form only over very warm ocean waters—at least 80 °F. For this reason, hurricane season in the United States tends to be at the end of the hot summer months, from June to November, when oceans near the equator are warmest.

Tropical depressions can build to tropical storms, which can build to hurricanes. A future hurricane first develops as a tropical depression, a clearly defined low-pressure system with winds below 38 mi/h. Some tropical depressions continue to build and become tropical storms, with lower air pressure and higher winds. Finally, a tropical storm that builds to wind speeds greater than 75 mi/h is considered a hurricane.

There are five categories of hurricane, with category 1 the least intense and category 5 the most intense. Hurricanes are ranked according to their wind speeds. Any storm with winds greater than 150 mi/h is considered a category 5 hurricane. Because hurricanes often move from water to land, they are accompanied by a surge in the ocean waters at the coastline. This wave of water, called a storm surge, can reach 20 ft or higher. Hurricanes are not to be confused with tornados. Tornados are violent rotating windstorms that form over land and affect a much smaller area.

Discuss the Formation of Hurricanes

[T] ➡ Display the transparency Anatomy of a Hurricane.
➡ As you discuss the formation of hurricanes, show some of the online satellite images and footage (optional).

Sample Questions

- How does a hurricane form? What conditions are necessary?
- What do the movies of hurricanes tell you about them?

Key Points

Tropical storms begin forming when a great deal of warm water evaporates into the atmosphere. This warm, moist air rises into the atmosphere, where a small change in temperature can cause condensation and cloud formation. Rapid condensation has several outcomes, the most significant of which is to dramatically lower the air pressure in the area. As all these water vapor molecules become liquid cloud droplets, the value of n drops dramatically, affecting the partial pressure of water vapor in the atmosphere. Ultimately, a hurricane ends up with very low pressure below and high pressure above. The pressure differential feeds the hurricane further, drawing more air into the storm. This cycle feeds on itself and is responsible for moving massive numbers of molecules, resulting in extraordinarily strong winds.

As the storm moves over areas of warmer water, evaporation increases. This also feeds the storm. A hurricane picks up wind speed as it moves over warmer water and loses wind speed over cooler water.

Water vapor density is related to temperature. As students saw in the previous lesson, the graph of these two variables is a curve that gets quite steep as the temperature increases. The steepness of this curve means that at warmer temperatures, small increases in temperature can have a dramatic effect on water vapor density. Thus, the warmer the temperature of the air and the ocean, the greater the chance that optimal conditions will exist for hurricane formation.

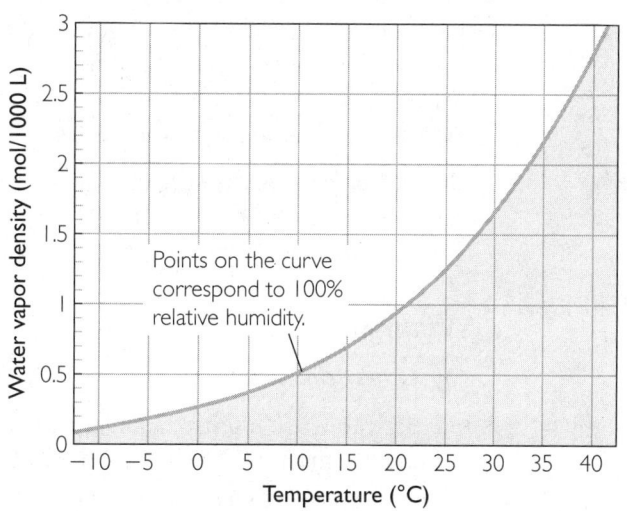

Water Vapor Density Versus Temperature

Points on the curve correspond to 100% relative humidity.

Discuss Global Warming (optional)

⇒ Display the transparency Global Warming Trends showing temperature changes on the planet over the past 120 years.

Global Temperature Changes (1880–2000)

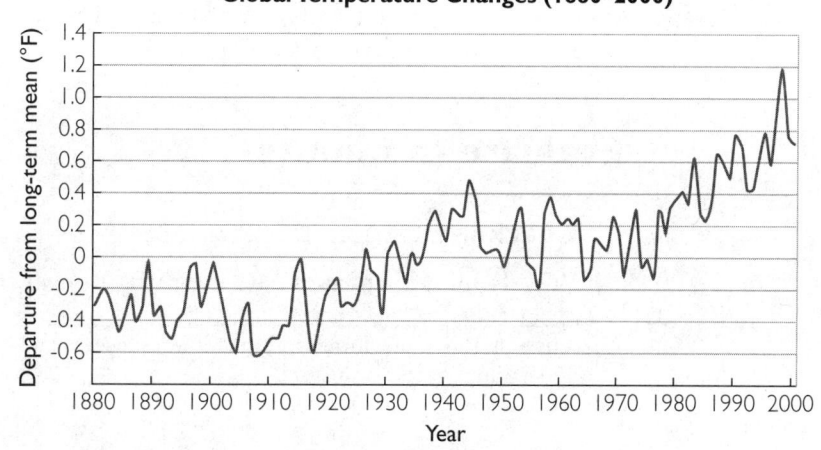

Source: U.S. National Climatic Data Center, 2001.

Sample Questions

- What is global warming?
- Why is global warming a potentially controversial subject?
- What can be done to try to slow down global warming?

Key Points

There is a great deal of scientific evidence supporting the idea that our planet is in a warming cycle, often referred to as global warming. Additionally, many experts speculate that the burning of so many petroleum products on our planet is a huge contributor to this warming. The practice of using gasoline, coal, heating oil, and so on as fuels releases enormous amounts of carbon dioxide gas into the atmosphere. The increase in carbon dioxide in our atmosphere has a "greenhouse effect," trapping heat from the Sun, and making it more difficult for the Earth to cool off. Global mean surface temperatures on the planet have increased 0.5–1.0 °F since the late nineteenth century. The twentieth century's 10 warmest years all occurred in the last 15 years of the century. Of these, 1998 was the warmest year on record, followed closely by 2005. A change of 1 °F may seem insignificant, but it has a dramatic effect on the water on the planet. Polar ice is melting at an accelerated rate, sea level has risen 4–8 in. over the past century, and both the number and the intensity of hurricanes have increased.

Global warming is causing climate change. Parts of the planet are experiencing wetter or drier weather than they have historically. Long-term climate change might mean more storms, floods in some areas, and droughts in other areas.

Wrap-up

Key Question: What are hurricanes and what causes them?

- Hurricanes are intense tropical weather systems accompanied by strong winds, massive amounts of rain, and ocean flooding. They form around low-pressure systems and are characterized by spiraling clouds and winds.
- Hurricanes form over the ocean in places where there is extremely warm, moist air.
- Rapid evaporation and subsequent condensation of moisture set up an air pressure differential that further feeds the evaporation and condensation.
- At higher temperatures, small changes in temperature have dramatic effects on water vapor density, n/V. Global warming may increase the frequency and intensity of hurricanes on the planet.

Evaluate (5 minutes)

Check-in

Why do most hurricanes have their origins near the equator?

Answer: Hurricanes form as a result of warm oceans and moist air, conditions prevalent in the ocean waters near the equator.

Homework

Assign the reading and exercises for Weather Lesson 19 in the student text.

Assign the Section III Summary to help students review for the quiz.

Assign the project Global Climate Change (optional).

LESSON 20 OVERVIEW

Stormy Weather
Unit Review

Lesson Type
Classwork: Pairs

Key Ideas
The weather is driven by the chemistry of phase changes and gas laws.

What Takes Place
In this lesson, students review what they have learned about phase changes and the gas laws and how these relate to the weather. They work in pairs on a review worksheet, then create a list of review topics for a unit exam. *Note:* This lesson is optional. You might instead want to use this time to have students present their projects to the class, or devote the time to the Unit Review in the student text.

Materials
- student worksheet

LESSON 20 GUIDE

Stormy Weather
Unit Review

Engage (5 minutes)

Key Question: What does chemistry have to do with weather?

ChemCatalyst

1. Describe three different physical changes involved in weather.
2. What physical changes of matter affect the weather?

Sample Answers: 1. Phase changes, changing temperature, changing density with temperature, changes in gas pressure, etc. 2. The phase changes of water cause water to cycle from the land to the air and back to the land. The expansion and contraction of air with changing temperature cause the air to move around the planet.

Discuss the ChemCatalyst

➠ Assist students in summing up how the topics they learned about relate to weather forecasting.

Sample Questions

- What does chemistry have to do with the weather?
- What causes matter to move on a global scale?
- How does changing temperature affect water? gases?

Explore (20 minutes)

Introduce the Classwork

➠ Pass out worksheets.
➠ Students can use any worksheets, notes, or handouts they have gathered over the course of the unit to complete today's classwork.

LESSON 20 CLASSWORK
Stormy Weather
Unit Review

Name _____
Date _____ Period _____

Purpose
To summarize and review what has been learned in the Weather unit.

1. Explain what happens at a cold front and why.

   At a cold front, a cold air mass is moving under a warm air mass. The warm air rises above the cold air because it is less dense. The cold front can cause short downpours.

2. Under what conditions does a front result in precipitation?

   The warm air must contain enough water vapor, and must cool enough when it rises to cause a phase change to liquid or solid water.

3. Why is rainfall measured in inches and not liters or milliliters?

   The volume of a container is proportional to its height in inches. Thus, the number of inches accurately reflects volume. By measuring in inches, the size of the rain gauge doesn't matter.

4. Describe how a liquid thermometer measures temperature.

   When a liquid is heated it expands; when it is cooled it contracts. Thus, the liquid in the thermometer will change level according to the temperature. Markings on the thermometer provide a scale.

5. Why does it usually snow more on mountaintops than at lower altitudes?

   The atmosphere is much colder up high, causing water to condense as it rises, increasing the likelihood of it freezing and forming snow.

6. Sample A contains 1.0 kg of rain, sample B contains 1.0 kg of solid ice, and sample C contains 1.0 kg of snow. Which sample will have the greatest volume? the least volume? Explain your reasoning.

   Each sample contains the same mass, but the density of rain is greater than the density of ice, and the density of ice is greater than the density of snow. So the rain will have the least volume and the snow will have the greatest volume.

7. Suppose you have 3 mol of helium in a balloon at STP. What are the temperature, pressure, and volume of the helium in the balloon? How many atoms of helium are in the balloon?

   The temperature and pressure at STP are 273 K and 1 atm. Because 1 mol of gas occupies 22.4 L at STP, 3 mol has a volume of 67.2 L. In 1 mol of helium gas there are 602,000,000,000,000,000,000,000 helium atoms, so 3 mol of helium contains 1,806,000,000,000,000,000,000,000 helium atoms.

8. A gas in a closed container with a movable piston has an initial pressure of 3.5 atm and a volume of 2.6 L. If the temperature doesn't change and the volume is changed to 3.4 L, what is the new pressure of the gas? What gas law did you use to determine the new pressure?

$k = PV = 3.5 \text{ atm} \cdot 2.6 \text{ L} = 9.1 \text{ L} \cdot \text{atm}$

$9.1 \text{ L} \cdot \text{atm} = P \cdot 3.4 \text{ L}$

$P = 2.7 \text{ atm}$

Boyle's law was used to determine the new pressure.

9. Find the pressure of 3.40 mol of gas if the gas temperature is 40 °C and the gas volume is 68.4 L. What gas law did you use to determine the new pressure?

$40 \text{ °C} + 273 = 313 \text{ K}$

$PV = nRT$

$P = \dfrac{nRT}{V} = \dfrac{(3.4 \text{ mol})(0.082 \text{ L} \cdot \text{atm/mol} \cdot \text{K})(313 \text{ K})}{(68.4 \text{ L})} = 1.3 \text{ atm}$

10. At the beginning of the day, a balloon containing 175 L of air is at a temperature of 78 °F. Later in the day, the volume of the air in the balloon increases to 186 L. What is the new Kelvin temperature of the air in the balloon if the pressure remains the same during the heating? What gas law did you use to determine the new temperature?

$78 \text{ °F} = 9/5(\text{°C}) + 32 = 26 \text{ °C} = 299 \text{ K}$

$k = V/T = 175 \text{ L}/299 \text{ K}$

$T = V/k = 186 \text{ L}/0.585 \text{ L/K} = 320 \text{ K}$

Charles's law was used to determine the new temperature.

Explain and Elaborate (25 minutes)

Review Phase Changes

Sample Questions

- What happens on a particulate level when a substance is heated? cooled?
- Why does hot air rise?
- What do phase changes have to do with the weather?

Key Points

Phase changes are a form of physical change. They are a central part of weather formation on our planet. Our planet is a watery one, with 70% of its surface covered with water in some form. Phase changes cause the transport of water from one part of the planet to another through evaporation and condensation. The Sun is the driving force in causing the water on the planet to change phase from a liquid to a gas. Once water vapor is in the atmosphere, it can easily move over the planet's surface.

Matter changes significantly when it is heated or cooled. When heated, the particles in a substance speed up, causing more collisions and increasing the average kinetic energy of the particles. If heated sufficiently, a substance can change phase, from a solid to a liquid, or from a liquid to a gas. The density of a substance also changes upon heating, usually decreasing because matter usually expands upon heating. When matter is cooled its particles move more slowly, and the average kinetic energy decreases. Matter usually contracts when cooled and its density becomes greater. Thermometers are designed to take advantage of the expansion and contraction of matter with temperature.

Review the Chemistry Covered in the Weather Unit

➭ Create a list of topics and terms on the board as a study guide for the exam. Encourage note taking.

Sample Questions

- What four variables are tracked for gases?
- What are the names and formulas of the gas laws? What must remain unchanged in the application of each gas law?
- How do you use the proportionality constant, *k*, in solving a gas problem?
- Describe a situation in which you would use each gas law.
- What is the Kelvin scale? Why is it important to use kelvins with the gas laws?
- How does the kinetic theory of gases explain temperature and pressure?
- What is Avogadro's law?
- What does the behavior of gases have to do with the weather?
- What is water vapor density, and how is it measured?

Key Points

It is necessary to understand the behavior of gases if you want to understand the weather. Scientists generally track four variables when they deal with gases: temperature, volume, pressure, and number of gas particles. These variables are all mathematically related. Knowing the mathematical relationship can help you solve problems associated with gases.

The temperature of a gas must be converted to the Kelvin scale in order to utilize the gas laws. For a specific gas sample, the proportionality constant, k, is always the same number. For a different gas sample, k is a different number.

Gas Laws

Charles's law $k = \dfrac{V}{T}$ **Gay-Lussac's law** $k = \dfrac{P}{T}$

Boyle's law $k = PV$ **Ideal gas law** $PV = nRT$ or $R = \dfrac{PV}{nT}$

Combined gas law $k = \dfrac{PV}{T}$ **Avogadro's law** $k = \dfrac{n}{V}$

The ideal gas law allows you to figure out problems that involve changes in the number of gas particles, n. The ideal gas law is represented by the equation $PV = nRT$, where R is a number that relates the different units to each other. For the units we have been using, $R = 0.082$ L · atm/mol · K.

Evaluate

There is no wrap-up or check-in for this lesson.

Homework

Assign the Unit Review in the student text to help students prepare for the unit exam.